MW00446101

HOW TO CREATE A
PROFITABLE BUSINESS
FROM YOUR SMALL FARM

THE FARM TO MARKET HANDBOOK

JANET HURST

Voyageur
Press

Dedicated to the small farmers I work with each day.
Thank you for being a constant source of inspiration.
Your efforts make the world a better place.

First published in 2014 by Voyageur Press, an imprint of Quarto Publishing Group USA Inc., 400 First Avenue North, Suite 400, Minneapolis, MN 55401 USA

© 2014 Quarto Publishing Group USA Inc.
Text and photography © 2014 Janet Hurst
All photographs are from the author's collection unless noted otherwise.

All rights reserved. With the exception of quoting brief passages for the purposes of review, no part of this publication may be reproduced without prior written permission from the Publisher.

The information in this book is true and complete to the best of our knowledge. All recommendations are made without any guarantee on the part of the author or Publisher, who also disclaims any liability incurred in connection with the use of this data or specific details.

We recognize, further, that some words, model names, and designations mentioned herein are the property of the trademark holder. We use them for identification purposes only. This is not an official publication.

Voyageur Press titles are also available at discounts in bulk quantity for industrial or sales-promotional use. For details write to Special Sales Manager at Quarto Publishing Group USA Inc., 400 First Avenue North, Suite 400, Minneapolis, MN 55401 USA.

To find out more about our books, visit us online at www.voyageurpress.com.

Library of Congress Cataloging-in-Publication Data

Hurst, Janet.
  The farm to market handbook : how to create a profitable business from your small farm / by Janet Hurst.
    pages cm
  How to create a profitable business from your small farm
  Includes bibliographical references and index.
  ISBN 978-0-7603-4660-0 (sc)
  1. Farm produce—United States—Marketing.
  2. Farms, Small—Economic aspects—United States. I. Title. II. Title: How to create a profitable business from your small farm.
  HD9005.H87 2015
  630.68'8--dc23
                                                    2014025098

Acquisitions Editor: Elizabeth Noll
Project Manager: Tracy Stanley
Design Manager: Cindy Laun
Cover Designer: Diana Boger

On the frontispiece: Maters and taters for sale at Wil Farm, outside of Hermann, Missouri. Roadside vegetable stands remain a popular type of operation for small farmers.

Printed in China

10 9 8 7 6 5 4 3 2 1

Disclaimer: This book is intended for education. It is not meant to take the place of professional advice from your tax accountant and attorney. No expressed or implied guarantee as to the effects of the recommendations can be given nor liability assumed.

CONTENTS

INTRODUCTION

For decades, farmers have been selling fruit, vegetables, eggs, butter, and other fresh produce at farm stands and markets. In the mid-twentieth century, such sales provided "egg money"—a little extra to buy a pair of shoes or a new coat.

When I was growing up in rural America, farmers' markets were a part of the portrait of the day. On Friday mornings, farm women would bring their goods to the courthouse square. From the back of a pickup truck, they would sell excess sweet corn, butter, and eggs. Some industrious women added homemade pies, cakes, breads, jams, and jellies. Other items such as colorful aprons, handwoven rag rugs and baskets, and handmade goods were often found.

In the 1940s and 1950s, these markets were simply a means of selling off extra products. The primary goods were eaten fresh, canned, or stored for family use. A little pocket money was gained from those Friday sales, and, in those days, quite a few pairs of shoes, a new coat, or other needed items were often paid for from "egg money." A handmade sign announced the market was, indeed, taking place; strategically placed boxes held extra vegetables, and pies were covered with a layer of plastic wrap. There was nothing fancy about these impromptu affairs. The traditional red-and-white tablecloth adorned a card table, a cigar box held some change, and then all that was needed were a few customers to make the day's sales begin. Customers were loyal, showing up each week, especially during strawberry, asparagus, and sweet corn seasons.

In the 1960s and 1970s, during the back-to-the-land movement, these little farm markets slowed down considerably. At that time, people who had an interest in growing food were busy on their own plot of land. These traditional Friday morning events fell out of fashion and remained that way until the late 1980s. At that time, a resurgence of interest in locally produced foods began to fuel a small fire. Markets began to come back and have been increasing in number since that time. The overwhelming popularity of these sales events makes a phenomenal statement: people want to know where their food comes from and who grows it.

FARMING TODAY

Today's market is not your grandmother's farmers' market. Now high-end, often elite, markets are found in neighborhoods, in cityscapes, and some still on court house grounds. Parks, businesses, and neighborhood pubs create market spaces, knowing future customers will come to buy and then enjoy their surroundings. Local is big business. Such national incentives as "Know Your Farmer, Know Your Food" have played a role in

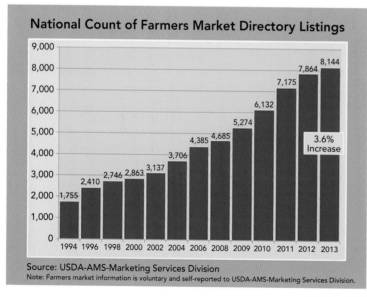

National Count of Farmers Market Directory Listings

Source: USDA-AMS-Marketing Services Division
Note: Farmers market information is voluntary and self-reported to USDA-AMS-Marketing Services Division.

USDA-AMS

8

bringing local foods to the forefront of the news and the marketplace.

Meanwhile back on the farm, the times they are a–changin'. Small farms are finding their way into bona fide food production and creating real income for growers. A force to be reckoned with, the small farmer is literally gaining ground each year. Backyards have turned into gardens, small acreage into mini farms, and large acreage into mega vegetable production.

With this burgeoning influx of farm goods, determining how to sell them can quickly become a problem. These products are perishable, after all. Perishable products don't offer a great deal of time for thinking, after the fact. Once the freezer is stocked and the excess is shared with family and friends, the thought process leads down the road of: "Could I make money from this?" The answer is yes. Currently, there is not enough local food to fill the demand. "According to census data, direct sales to consumers account for a small but growing segment of U.S. agriculture. Direct sales of agricultural products, including

Could I make money from this? The answer is yes.

Internet and catalog sales, totaled $1.2 billion, or 0.4 percent of total agricultural sales in 2007, or 0.4 percent of total agricultural sales. From 1997 to 2007, direct sales increased by 105 percent ($619 million), compared with an increase of 48 percent for all agricultural sales ($95.8 billion). Over the same period, the number of farms selling directly to consumers increased by 24 percent, compared with a 0.5 percent reduction in the total number of farms. In 2007, the 136,800 farms selling directly to consumers surpassed the number performing custom work (services for others, such as planting, plowing, spraying, and harvesting), agritourism, and sales of firewood and other forest products, positioning direct sales as the leading on-farm supplemental enterprise."[1]

Yes, this is great news for the small farmer. Yet the problem of getting products from point A (the farm) to point B (the customer) still exists, even though there are a variety of means to get the food from the farm to the market. We all learned in Boy Scouts, Girl Scouts, or 4-H to "be prepared." First develop a marketing plan before actual farming even begins. One of my mentors gave me this advice: "Before you plant your first seed, know who is going to buy your crop." I've passed this advice on to others, and many plan their seasons with that practice in mind.

EXERCISES

The exercises in this book were developed to assist you in a number of areas:

- Developing a marketing plan. Once this tool is created, it is a fluid document, meant to guide and direct the path of the business. If you have not considered the farm as a business, it is time to do so.
- Exploring the available markets. Using the marketing plan as a guide, we will take a look at the pros and cons of methods of direct sales.
- Learning from others' experiences. We will interview producers in various stages of business development, learning from their successes and mistakes.
- Exploring which farmers' markets work. We will examine some of the most successful farmers' markets in the United States.
- Exploring other ways of marketing. Farmers' markets aren't the only means of marketing. Other methods of marketing will be explored, including community supported agriculture (CSA) and CSA development.
- Obtaining organic certification.
- Becoming familiar with Farm to School programs.
- Learning about on-farm sales, such as farm stands, u-pick, and traditional retail outlets.
- Working with a broker.
- Knowing what it takes to be in business, including food safety, Good Agricultural Practices (GAP), insurance, regulations, and electronic benefits transfer (EBT).
- Developing a good attitude—putting the best food forward, customer service, and professionalism.

OPPOSITE: The heart of the farm: A circa 1890s barn awaiting new arrivals. Note the hayloft door and two-over-two windows. Sunflowers are established for wildlife.

1 "Consumer Demand for Local Food Increasing," Organic Guide. November 18, 2010. www.organicguide.com/organic/news/consumer-demand-for-local-food-increasing/.

CHAPTER 1

ELEMENTS OF MARKETING

Geese at Gosherd Valley. Courtesy of Connie Cunningham

Lisa Murphy of Sweet Spirits Farm, in Hannibal, Missouri, proudly displays her line of handmade goat milk soaps.

It has often been said that there is nothing new under the sun. While this is largely true, there *is* something new going on in the world of agriculture. There has been an awakening, a re-creation within the farming movement. It began with people raising a few chickens in their backyard and turning suburban lawns into vegetable gardens. America is waking up to the need to produce and consume high-quality food.

As with every movement, there are those who wish to benefit from their skills and earn extra income from their efforts. America is still the land of opportunity! If you are reading this book, let's assume your dream centers upon running a small farm and generating some income from that farm. In days gone by, *farm* typically meant a hundred or more acres, tractors, combines, corn, beans, and a lot of cows.

In the past few years, the term *farm* has come to have many different meanings from an urban section of land that is cultivated and brought into vegetable production, small plots of rural land producing vegetables or animal protein, and midsize parcels of land and large landholdings. Each of these types of operation can fall under the term *farm*.

SMALL FARMS

Many sizes and shapes of farmers and farms exist in the United States. It is encouraging to know, however, that a person can still fulfill a desire to be a farmer without all the bells and whistles of modern-day, large-scale agriculture. Shovels, rakes, and hand tools still function as they did a hundred years ago, as long as there is a willing victim to operate these tools.

Farms can be large or small. We still see the mega farms producing primarily corn and soybeans. Other farms concentrate on wheat. There are large dairies, cattle operations, and hog operations across the United States. However, the small farm is gaining new attention. These farms can range in size from backyards to several acres.

The US Department of Agriculture (USDA) defines a small farm as "One that grows and sells between $1,000 and $250,000 per year in agricultural products." By that definition, there are more than 2.1 million small farms here in the United States.[1] Incidentally, women are coming on strong as farmers. Nearly 1 million women now claim farming as their primary occupation. Minority farmers are also making their mark. According to the 2007 National Agricultural Statistics Service (NASS.usda.gov), there were 66,671 farms with Hispanic, Spanish, or Latino operators; 61,474 Alaska native or American Indian operators; and 32,938 African American operators.[2] A statistic from the Bureau of Labor, which is eye-opening, is that the average age of farmers, as a whole, is nearly sixty years old. Do you find that troubling? Who will feed the nation? If there was ever a time for the small farm movement, it is now.

One area that has come on strong is the urban farm. These small plots are often reclaimed land in the heart of the city left vacant by deconstruction. Through intensive planting, raised beds, and soil mitigation, city lots, which were once eye sores, are becoming productive mini farms. Thousands of pounds of food are coming from these parcels of land. Though production space is limited, creative planning and planting brings surprisingly large yields. These farms serve as a glimmer of hope in neighborhoods plagued by violence and lack. Neighbors begin to come together to assist or claim a space in a community plot. Who would believe vegetables could create an urban renaissance? It is possible and it is happening in cities all across America. Private groups, church groups,

If there was ever a time for the small farm movement, it is now.

not-for–profits, and other organizations are reclaiming land with the support of city officials. They, too, would rather see this land in production rather than as a dumping ground for refuse.

If you have an interest in starting an urban farm, begin with your city planning and zoning commission to find lots suitable for this type of project. Organize neighbors and friends. Since many of these lots are former housing sites, soil tests are very important. Most who farm in an urban setting bring in soil, use raised beds, or use other sorts of growing media.

Don't be surprised to see a few chickens, a beehive, or maybe even a goat or two on these urban parcels. Variances are sometimes granted for these special circumstances.

Recent trends tracing the literal roots of food back to the producer have given small farms a new breath of life. No longer does a producer have to grow and process thousands of chickens a week or grow entire fields of one particular crop to see an increase in income. Consumers are learning the true value of good food, and those consumers with average or above average incomes are willing to part with a significant portion of their grocery budget to pay for high-quality fresh goods. In the eyes of the educated food buyer, bigger agriculture is not better agriculture. Premium prices are paid to "know the farmer," which is a refreshing trend. This new focus is driven by sustainability.

The term *sustainable* has become closely identified with small farmers. According to US Code Title 7, Section 3103, "The term 'sustainable agriculture' means an integrated system of plant and animal production practices having a site-specific application that will, over the long-term—

(A) satisfy human food and fiber needs;
(B) enhance environmental quality and the natural resource base upon which the agriculture economy depends;
(C) make the most efficient use of nonrenewable resources and on-farm resources and integrate, where appropriate, natural biological cycles and controls;
(D) sustain the economic viability of farm operations; and
(E) enhance the quality of life for farmers and society as a whole."[3]

That definition is a central element of the legislation of the Sustainable Agriculture Research and Education (SARE) program of the National Institute of Food and Agriculture (NIFA).

MARKETING

A farmer must wear many hats: horticulturalist, animal husbandry expert, mechanic, bookkeeper, peacemaker, and marketer. Marketing has become a necessary part of deriving an income from a farm. Farmers have become savvy in this arena and have learned that sales equals income and that sales are all about eye appeal, market development, and catering to the buying public. It does little good to raise twenty-five pounds of salad greens and have nowhere to go with them. Planning is key.

There are some services worth paying for. It is important to have expert advice in legal matters, accounting and tax preparation, and insurance coverage. Include these professionals as a part of your planning team.

So what are the elements of a marketing plan? The basics are the same whether you are selling socks, lollypops, or green beans. Of course, the rules about handling products vary, but marketing remains the

DEALING WITH LABOR SHORTAGES

Finding laborers is often one of the most difficult obstacles for a farmer to overcome, but there are means to assist with this ongoing problem. Farmers often bring apprentices onto the farm. Although they are usually not paid, apprentices can learn valuable skills from the experienced farmer while receiving room and board. Of course, bringing someone onto the farm creates an additional responsibility for the farmer, but many growers find this responsibility worth the exchange in labor. Some industrious farmers actually charge a fee for an apprentice to work with them.

Occasionally there are volunteers who want to come and experience the farm environment. One local farmer had a neighbor who was dealing with a loss in her family. She asked to come to the farm and pick green beans. She simply wanted to do something worthwhile that got her out in the field and occupied her hands and mind. She became such a fixture that the farm family nicknamed her Bottom's Up! It turned out to be a great deal for both parties. Bottom's Up went home with a share of green beans for her efforts, not to mention a lighter heart, and the farm family had buckets of green beans to show for her labor.

Worldwide Opportunities on Organic Farms (WWOOF; www.wwoofusa.org) is a nonprofit network of host farmers and volunteers who wish to connect to work out a labor exchange. Its website states: "Since 2001 we have been connecting sustainable farmers with willing volunteers, in an exchange of education, culture, and sweat to bring forth wholesome agricultural products from the farms of the USA." Visitors, called WWOOFers, are from varied backgrounds and volunteer on farms to receive some practical experience and learn about farming and organic agriculture. Labor is in exchange for room and board. These arrangements seem to work out very well; once farmers have a successful experience with one WWOOFer, they are anxious to open the door to another.

Internships are another way of involving people in your farm. You have learned valuable things over the years. Some may seem elementary to you. However, there are people willing to work to gain knowledge. Consider offering an internship (paid or unpaid) on your farm. Plan to spend time with the intern, actually teaching skills and sharing your knowledge.

While sharecropping is not as prevalent as it once was, this is also a means to assist with labor shortages. Workers literally labor for a share of the pickings. This can be a good value for both the worker and the farmer.

same: the satisfaction of a need or want fulfilled. An overall marketing study would look like this (please see Worksheet 2, Appendix 1):

- Project background
- Proposed operating structure
- Availability of supply
- Labeling
- Description of products to be marketed
- Expected pricing
- Competition
- Processing
- Food safety
- Marketing trends
- Distribution
- Marketing tactics
- Execution of marketing strategy

We will develop these points in the next chapter. While constructing such a plan may seem like a time-consuming task, defining your route can provide a clear course to follow. On days when the debate is on whether you should go in one direction or the other, or take on this project or that one, it is worthwhile to have a document to review and ask, "Does this fit in with my existing marketing plan?" The plan will provide an interactive road map to assist the small farmer in the overall production and end sales from the farm.

If you are not a writer and don't enjoy this process, don't despair; a marketing plan can be as simple or as complex as you choose to make it. Your first impression may be to discount all this information and simply wing it.

Before you buy your first bee, know where you are going to sell your honey.

However, it has been proven over time that "those who fail to plan, plan to fail." If farming is a long-term goal and if it is to be a money-making proposition, then some planning sessions will be time well spent.

SWOT ANALYSIS

Perhaps the very best place to start is with your vision for your farm, or, more accurately, your farm business. What are your goals? Do you want to add $1,000 a year to your income from a small garden spot? Do you want to add $5,000 a year from a specialty crop or $10,000 from extended season growing? Let this percolate a while, and then establish a realistic goal for year one. This is a good time to introduce a tool specifically for this purpose: SWOT. SWOT analysis focuses on four areas (see Worksheet 1 in Appendix 1):

- Strengths: What are the strengths of your proposed business? For example, are you organic? Do you have the capability to extend your growing season? Do you have a specialty product? Do you have supportive family members? Extra hands come harvest time? Do you have a packing shed? Anything that can be termed a strength should be written in this section.

- Weaknesses: What will cause you trouble? Not enough help? Not enough land? Not enough equipment? A growing season that's too short? Record anything that will be a hindrance to your proposed operation. A recent poll at a growers' conference showed lack of available labor as one of the primary obstacles for many farmers to overcome.

- Opportunities: What can you do to expand your horizons? Is there a niche

Alan Nolte of Nolte Hills Farm, Morrison, Missouri

"My family was in the nursery business for over 35 years. When the economy tanked, our contracts for installations on new construction projects came to an abrupt halt. With fewer houses being built, our trees and shrubs were no longer a viable income source. We started planting vegetables and over the past five years, we've had a steep learning curve, but the vegetables saved our business and took us down a whole new road. I always wanted to be a farmer."

market you can fill? Are you particularly known for a specific product? Are there marketing tactics you can employ to build a better mousetrap?

- Threats: What if you as the primary operator are unable to fulfill your intended role? What if insect damage destroys your crops? What if a fox comes in and eats all the chickens? Nature itself proposes many threats! Farming is not for the faint of heart. In this section, record the foreseeable problems.

Writing a business plan, as you'll do in this book, is intended to open your eyes.

Some folks may approach farming as a pie-in-the-sky, going-to-make-a-million-quick kind of endeavor. With all the attention on local foods, there will be those who join on a whim. Reality has a way of setting in quickly. There is no easy route to a successful farming operation.

FARMING AS A LIFESTYLE

There also has to be a buy-in; farming is not only a means to make a living, but also a lifestyle. Life on a small farm can involve the whole family. Children grow up with chores, learning responsibility, and families learn to share in times of bounty and in times of lack. Farm kids are proven to be a resilient, resourceful lot. That in itself is a legacy.

Beyond that, the ability to plant a seed and watch it grow into food, to learn to milk a goat, to truly understand the value of life, and the bonds between humans and animals are life lessons rarely learned in our plastic, wired world.

Keep in mind that there are certain intrinsic values in farming that cannot be noted on a spreadsheet or in a business plan. That being said, farmers are famous for working for wages well below the poverty level. If you are figuring prices, it is imperative to calculate a living wage. Otherwise, all your efforts are for naught. Figure your time investment in the final cost of goods to be sold. All those trips to pick up supplies, to deliver, those early mornings and late night

The University of Missouri Extension offers a course in writing a business plan. Grow Your Farm focuses on writing a business plan for farming with marketing as a segment of the overall strategy. Take a look at the four *p*'s of marketing:

1. Product: Consider different aspects of products, including quality standards and how outside influences might affect production and sales.

2. Price: The price determines profit. The key to setting a price is to know your product and customers. Producers have been price takers in traditional markets. You have the opportunity to be a price maker in nontraditional and value-added markets.

3. Promotion: Advertising is only one aspect of promotion. Promoting products is more complex than ever, but today's diverse markets offer many opportunities. How well you promote your products will determine whether you stay in business.

4. Place: Where you are and where your customers are will influence many of your business decisions. Place relates to where you will sell or market your product and the distribution channels you will need or have available to get your product to your customers. Place influences the middlemen you will have to involve.[4]

hours add up. Calculate those hours and average them out over the season to come up with a final figure.

Farming for a living may seem to be an ideal way of life, but it is, in truth, fraught with challenges and difficulties. Small irritants, such as gravel roads and power outages, can wear on a person over time. Are you the sort who can raise an animal for slaughter? Some people can and others simply cannot raise an animal and then see it on the table as dinner. There are many decisions to make before taking on rural living. Talk to folks who have taken the plunge and ask for honest answers to your questions. Assess your own values and those of your spouse or partner. It is very important to have all the oars in the water, rowing in the same direction!

Perhaps this is the point where you ask yourself, "Why do I want to farm?" Is it for independence, the lifestyle, the family involvement? Is it because producing food is a trend, because your friends are farming? Is it what your family is known for? There are as many reasons to farm as there are vegetables in the field.

What are your assets? Do you already own land? If you want to purchase land, do you have the cash or can you obtain credit to do so? If you borrow, can you afford the associated payments? Can you rent land or make other arrangements to farm? (See Worksheet 4, Appendix 1.)

Do you live on the land or do you have to travel to it? All these things factor into long-term decisions. If you don't live on the land, if you want to have livestock, such as chickens or goats, you will have to make arrangements for an enclosed chicken yard and/or a guard animal for goats. The majority of livestock requires daily care.

LEARN ABOUT FARMING

With the newfound interest in food production of all sorts, educational opportunities for farmers abound. Land-grant universities (See Appendix) offer Extension services, bringing the university directly into the field. Check with these institutions to learn about workshops in everything from butchering to seed saving. This is the place to learn from the experts and receive the latest research-based information.

Civic organizations and private foundations have jumped on the food bandwagon as well. Gardening and canning classes, pest identification courses, almost every imaginable topic can be found through botanical gardens, herb societies, and master gardeners organizations.

The old adage to find a need and fill it has always been at the heart of marketing. In Chapter 2, we will work on the marketing plan as a whole.

1 "Demographics," US Environmental Protection Agency Ag 101. Last updated on 4/14/2013. www.epa.gov/agriculture/ag101/demographics.html.

2 "Farms with Women Principal Operators Compared with All Farms," 2007 Census of Agriculture: United States. Accessed May 29, 2014. www.agcensus.usda.gov/Publications/2007/Online_Highlights/Race,_Ethnicity_and_Gender_Profiles/reg99000.pdf.

3 "United States Code, 2011 Edition Title 7 Agriculture," US Government Printing Office. Accessed May 29, 2014. www.gpo.gov/fdsys/pkg/USCODE-2011-title7/html/USCODE-2011-title7-chap64.htm.

4 Grow Your Farm 6-2; University of Missouri

YOUR MARKETING PLAN

What will you sell? Where will you sell it? Who are your competitors? How much space will you need? A marketing plan helps you figure out the details.

DEVELOPING THE PLAN

Now on to the crux of a plan. See Worksheet 2 in Appendix 1 for a template of a marketing plan. There are shorter and longer versions of marketing plans available. This one is thorough, so it is a good choice for a comprehensive overview of a new or existing business. An overall marketing study would have the following elements:

- **Project background:** Explain the premise of your business. How do you define yourself? Are you a CSA, a small farm, a co-op, a partnership, or another type of organization? This is the place to tell your story, to give your elevator speech. Who are you and what do you do?

- **Proposed operating structure:** Who is doing the work and what is their background? Often people think they need formal instruction to apply to this section. While education is great and should be noted, practical experience should also be applied to this section. Have you participated in other organizations? What business or growing background do you bring to the project? Who will be involved and what are their strengths, experiences?

- **Availability of supply:** How much product is anticipated? This is the million dollar question! It is hard to know if it will be a year of glut or lack. Either is not the prime placement within the market. This is where a simple educated guess will apply. If you have records from previous years, so much the better; but if you are a new farmer, it is simply an educated guess. The answer to this question will drive your market and tell you where you need to go to market your products. Scale is everything.

- **Labeling:** Depending on your product this may or may not be necessary. If you are legal to produce baked goods or jams and jellies in your market area (city or county), then the label will require your name, address, phone number, and name of the farm, if applicable. Then a listing of ingredients in the order of volume (for cherry jelly: cherries, sugar, pectin; for cinnamon bread: flour, sugar, eggs, milk, sugar, yeast, cinnamon, etc.).

  Labels for meat and eggs are specific to the product, so check with the county health inspector to find out the guidelines for these specific products. USDA approval is required for meat, and an egg license is required for eggs. (Egg licenses are available through your state's Department of Agriculture.)

  Some retail outlets will also require products to carry a UPC code. Ask the store about the format for the code. The store should be able to assist you in obtaining the proper code. This will be your unique series of numbers, which should be transferable among various retail outlets.

- **Description of products to be marketed:** This is your anticipated inventory. What will you grow or produce? Take a look at your product mix. Will you be known for one thing or will you offer a variety of goods? There are advantages to both.

- **Expected pricing:** This will fluctuate depending on your product, the season, and availability. There are various means of arriving at a price. Terminal prices are available throughout the United States for vegetable pricing. The USDA Economic Research Service[1] offers a listing of average retail prices for common vegetables.

Terminal prices are also available through the USDA Agricultural Marketing Service: "AMS services for the US produce industry help buyers and sellers market their perishable products in the most efficient manner possible through distribution channels. We partner with state agencies for the benefit of nationwide growers, shippers, brokers, receivers, processors and the foodservice industry."[2] Market pricing is also available for beef, pork, lamb, and goat via the USDA Economic Research Service.[3]

- **Competition:** Who are your competitors? Take a look at retail outlets, grocery stores, big-box stores, and other farmers' markets to see who and what your competitors have to offer.

- **Processing:** This may or may not be necessary depending upon your product.

- **Food safety:** See Chapter 10: Food safety

- **Marketing trends:** An overview of the existing market. Read headlines to find out what's new in the food market, note trends in recipes, and pay attention to stories about foods with added health benefits. For example, who knew kale would be the rock star vegetable of the future? It is hard to pick up a magazine these days that has a recipe for a dish that doesn't include kale. Trends sell food. Heirloom vegetables often bring 1.5 times the price of hybrids. Let your customers know what you have and why it is special. Remember people want to know where their food comes from and feel

good about it. Bring some pictures of your farm to give yourself talking points. Involve the customer in what you do.

- **Distribution:** How will you deliver the goods? Is refrigeration necessary? How will you handle this? Do you have a refrigerated truck? Can you get by with coolers? Does the marketplace offer a place to plug in a refrigeration unit? All questions are to be considered. Now's the time to start thinking about your retail outlet: will you sell to an established grocer, a co-op, direct market, or all of the above? Volume will be your guide to these questions.

- **Marketing tactics:** This might include signage, displays, brochures, recipes, overall branding, or identification of the target market. Okay, here's where the concept of branding comes in. You are creating an identity for your farm. Design a logo, come up with a consistent text format, one that you will use on everything that has anything to do with your farm. The single most important thing to give your customers in the way of advertising is a business card. It immediately establishes you as a professional and gives the customer something to take

Design a logo and use it on everything.

home. Many farmers relate their success of repeat sales to a simple business card. From there, go on to what you can afford, such as stickers, brochures, recipes, etc. Always, always put your name and contact information on anything you hand out. If you have a booth at a market, then hang a sign so people come to know their farmer.

- **Execution of marketing strategy:** This is how all of the above will come together.

Yes, at this moment it probably sounds overwhelming. If you look at this type of endeavor as a whole, you may want to quit before you begin. However, like all monumental tasks, if you break it down into steps, points on a scale, then it will start to make sense and come together bit by bit.

As you work through the development of your marketing plan, make sure to include partners, either business or life, or those closely involved in your farm. Moving

Liz Graznak, of Happy Hollow Farm, in Jamestown, Missouri, sells her vegetables at the Columbia Farmers' Market. A farmers' market is just one place to sell your produce—you can also sell to chefs, schools, brokers, and more.

ahead, we'll take a look at farmers' markets, CSAs, co-ops, and other types of farm-related businesses. There may be some that you have not even considered.

PUTTING THE PLAN TO WORK

Review your marketing plan with your team. (See Worksheet 3, Appendix 1, for help figuring out who your team is.) It is important to have everyone on board and moving in the same direction. After the plan is written, there will be a document to go to, to assist in decision-making processes. Although nothing is set in stone, proceed with caution if there is a shift in the overall objectives of the business, and make sure that the shift does not divert the farm from the overall goal of making a profit. This may mean letting go of something, and that can be hard to do. It is important to lead with the head and not the heart in these matters.

THE MARKETS

With this information in hand, you are ready to take a look at the various types of markets and how they can work for you. Traditional marketing routes include:

- **Farmers' Markets:** These are often community-sponsored events that usually take place one or two days a week. Participating in farmers' markets is a good way to begin to market your products. Many farmers prefer this method for several reasons, number one being getting instant feedback—a little praise goes a long way. If customers are excited, farmers' markets can fuel the grower's fire. Cash sales are another benefit of farmers' markets. Many customers come prepared with cash when they visit a market. Cash on the barrel head is still a good way of

doing business. Farmers also develop relationships and build trust with their customers at farmers' markets.

- **Community Supported Agriculture:** CSA is a method of marketing that has customers buying "shares" of the anticipated crops. Then each week throughout the growing season, they are given a box full of vegetables. The box contains vegetables that are available that week, some weeks will contain more, others less. The idea is that the customer is sharing some of the risk of farming. If it is a bountiful year, the customers will receive a greater amount of goods. If it is a dry year, then they will receive less. The advantage to this type of arrangement is that the customers pay upfront for the season, so operating capital is at hand when it is most needed. The downside can be that during a lean year, it is hard to get customers to buy into the program a second time. It is also a little hard to have all the payments upfront and most likely spent, while working the rest of the summer. Even though payment has been made, sometimes it is good to have rewards coming in along the way. Those who sell in this manner work for years to develop a loyal following and find their CSA to be a necessary, valuable source of income for their farm.

- **Brokers:** Brokers can represent your farm, selling your produce for a commission. Often, these people have developed relationships with chefs or retail outlets. They have a customer base to work from, which is a huge help to new farmers. Chances are, the farmer has little time to go out and develop these relationships, especially in the first few seasons of starting up the operation. A broker has the inroads

People want to know where their food comes from and feel good about it. Let them know what is special about your product.

already established and if his or her reputation is good, then this can open many doors that would otherwise be difficult to access. Of course, there is a fee for their service. Sometimes, the broker buys the goods from the farm and then simply marks them up. This is a good way to do business, with the farmer setting the price first. Remember to be a price maker, not a price taker, if at all possible. This type of arrangement works well for chefs because they can tell the broker what they are looking for and avoid a parade of farmers entering their kitchens. Brokers work particularly well for those who would rather be on the farm than entering into the sales arena. For many farmers, selling is their weakest point. A broker will handle the transactions and chances are that selling is his or her strong suit.

- **Co-ops:** Co-ops were a popular idea in the 1970s with their small storefront operations working in harmony to provide vending space for multiple sellers. This concept allowed the vendors to share expenses, such as rent, utilities, advertising, etc., while providing a market space for the members. Work share agreements allowed the vendors to be on the sales floor for a day or two a week, while other members worked the other days. This type of arrangement can still work today as long as there is a firm understanding (documented) of who will handle the day-to-day operations, how bills will be paid, how farmers will be reimbursed, etc. A co-op arrangement can be a stretch if this type of business is with family or close friends. Remember to keep business separate from your personal life and have a structured method of operation.

- **Traditional retail outlets:** Some farmers prefer to deal directly with produce managers in larger supermarkets. While this may be possible for those growers with a large volume of production, often the small farmer is excluded from this market due to lack of available goods. There are also regulations to consider. Most large stores ask for liability insurance and food safety plans. The store must add its markup, typically 40 to 100 percent, to your price. Keep this in mind when considering this option.

- ***Farm to School programs:*** This seems like a no-brainer. We want healthy children, so the best medicine we can give them is healthy, nutritious foods. If we feed them chemically laden food, then we will pay for this decision with extra medical bills and more health concerns.

School systems are coming on board with the Farm to School program, and there is growth in this industry annually. Perhaps selling your produce to the area school system will open a new market for you. Set up a time to visit with school administrators. Have your ducks in a row and present your ideas in a confident, businesslike manner. Take examples of Farm to School program material throughout the country for review. Start a conversation at the PTA, see where it all goes.

- **On-farm sales:** Many farmers prefer on-farm sales, such as farm stands and u-picks. Buyers often enjoy a trip to the farm as

continued on page 32

Heirloom varieties of squash turn up early at the market, thanks to high tunnel innovation.

Todd Geisert

Todd Geisert of Geisert Farms (www.toadspigs.com).

Todd is the fifth generation of his family to farm a piece of land an hour outside of St. Louis, Missouri. The Geisert family has been raising hogs since 1916, and little has changed on the farm since the early days. Todd's father, D. John, built A-frame huts to shelter the hogs; Todd uses recycled signs and billboards to construct his huts. The huts are moveable, so the animals can be raised in different places on the farm, making use of various pastures that were used the previous season to grow corn or soybeans. Modern equipment is now used, instead of horses, but largely, techniques are the same as they were in the past. You may ask what keeps a nearly 100-year-old farm viable. According to Todd, it's a commitment to doing things the "right way" while having an eye on the future: combining good old-fashioned animal husbandry techniques with modern technology. Todd's slogan is "Doing it the natural way," and that certainly applies to his farming methods—but at the same time, he's rarely seen without his iPad.

There's another reason Geisert Farms (www.toadspigs.com) is so successful: Todd has created a diverse market for his pork products. He sells pork and other produce from a farm stand, he sells to area chefs and butchers, and he hosts agritourism events.

Todd's location benefits the farm in many ways—it's rural, but still it's close enough to the city to work with urban chefs. Todd has worked diligently to develop working relationships with local chefs, who appreciate the fact that he raises hormone- and antibiotic-free pork. Todd's pork is often featured on area menus, and he is well known in culinary circles. It is not uncommon for Todd to be in jeans by day and dress clothes by evening, circulating between the farm and area restaurateurs.

Todd also works with local butchers, supplying various cuts of meat, cured meats, and snack sticks. Nothing is finer than a whole hog roasted from the farm!

Todd hosts "Chefs in the Pasture" events and invites area chefs to come directly to the farm and see how the pork is raised. Local wineries and breweries also participate in this event. Todd's pork burgers are known far and wide as the best. "They fly off the grill as quick as we can cook them," states Todd. Todd's wife, Katie, is active in the marketing of products and supplies samples for events.

In addition to the full line of pork products, Todd raises vegetables. Most of the vegetables are sold in his self-service farm stand. There people can shop for meat, vegetables, melons, and other produce items. It is not uncommon to see cars lined up on weekend days, waiting for a chance to load up on pork burgers and other favorites. Customers leave payment in a box, and Todd says he has never had a problem with the honor system.

Todd has crossed the bridge from country farmer to marketer extraordinaire. His sincere manner and his cross-marketing efforts bring people to the farm and allow the farm to be successful in a time when family farms are rare.

Goats await milking time at Goatsbeard Farm.

continued from page 29

well. However, it is important to consider the day's schedule and the volume of customers. It is important to make the customer feel welcome, not as an interruption to your work. Many small farm stands still work on an honor system. This way the customer can come, shop, pay their bill, and take their produce with them, all on their own. Surprisingly, those who operate in this manner report very little loss. It is encouraging to think that commerce can be conducted in such a simple manner in this day and age. The farmer supplies a calculator, sales pad, money box, bags, etc. Customers record their purchases and leave their money in the box. The downside to this method of doing business is that it requires payment by cash or check, and customers must think a little bit ahead to prepare for shopping. However, after the first time, customers will understand how the system works and will be prepared with cash or check. If the farm is on a busy route, this is a good fit for a farm stand. Incidentally, this is the highest area of profit because little to no travel is involved and the stand is unmanned. However, some initial investment will be required in the stand, refrigeration, display tables, etc.

- **U-picks:** U-picks have gained in popularity over the past few years. Berries are often sold in this manner, allowing the customer the experience of picking fresh berries and filling a bucket provided by the farmer. Fresh asparagus is sometimes sold this way with customers being given a marked stick to show them how high the spear should be for picking. Of course apples are a favorite fall crop, and who can forget a trip to the orchard to pick a

Goatsbeard Farm has been making artisan cheese for more than 20 years. Their market is divided between direct sales and wholesale.

bushel? Many entrepreneurs take advantage of their audience, adding items available for purchase, such as hot cider, donuts, apple butter, and other farm favorites. How about inviting a photographer to come and take family photos in the orchard? Customers love destinations and the opportunity to create a memory with family.

- **Working directly with chefs:** As already mentioned, developing a working relationship with chefs can take some time. First of all, there are others already supplying the restaurant with goods. Growing something unique or in high demand can help you open the door, but reaching chefs is not an easy task. They are busy and their schedule must come first. Be prepared to meet them on the run and give them your elevator speech. Introductions from a broker or a shared acquaintance is a good way to meet them. Ask what they need that they cannot find. If you come to an agreement, take only your best goods and cultivate the relationship as carefully as you do your crop. Bring them samples of other things you have to offer. Call them if you have something special. Go the extra mile to make these relationships fruitful for both parties.

- **Internet sales:** A great deal of commerce takes place without ever meeting the customer. Internet sales have become a huge vehicle for the specialty foods market. If you produce an unusual product, then consider the Internet as your sales floor. With the advent of overnight shipping, specialty packaging, and other progressions in food

delivery, this could be the goose that laid the golden egg.

- **Produce auctions:** These are regional auctions specifically for the sale of produce, plants, and other goods.

- **Sale of services:** Turn your passion of farming and gardening into a business. Many people want a beautiful garden but simply don't have the know-how. Offer your services for a fee: assist with the building, planting, and tending of raised beds or other garden sites. The customer will reap the rewards.

- **Rent a goat:** Don't laugh! In many areas, goats are employed as brush tenders or lawn mowers. Consider renting out your goats as weed-control agents. Airports and parks across the nation are renting animals for this purpose. Along the same line of thinking, consider a mobile petting zoo for fairs and festivals. There is quite a bit of work in the setup, so charge accordingly. Check with your insurance provider, too.

- **Write about what you know:** There are hundreds of magazines, many for homesteaders, gardeners, and small farmers. If

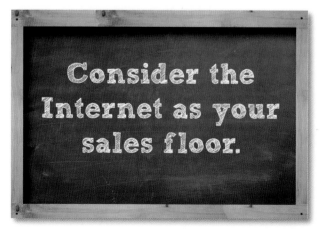

Consider the Internet as your sales floor.

you have knowledge of a specific area and the gift of spinning a yarn, write about what you know and sell articles to consumer magazines. Once you get started, there is nothing like seeing your name in print! Do you have a specialty? Do you know something your grandmother taught you to do that not many other people know (a special bread, a special tea, etc.)? Write about it and submit your article to a magazine you like to read. Start keeping a rolling list of your printed articles and attach it to every article you submit. Before long, you will have a list of several articles, and that lends credibility. Remember to target your audience and write for that particular sector of the population. Consider interests, age, gender, locale, and other factors when writing for a publication. Chances are you have a wealth of knowledge you haven't even considered yet.

- **Gift baskets:** Gift baskets composed of locally produced foods are popular. This format works well for those who live in areas where tourism is an active part of the economy and also in locations (such as farmers' markets and other venues) where localism is a priority.

There are many different types of marketing opportunities. Much of the choice

Sample baskets are filled and ready for customers. Check your county and city codes for rules about selling baked goods (and offering samples).

of marketing venues will depend upon a farmer's style, comfort zone, and location. If the farm is located in a rural area, chances are there will be some travel involved. Find the best match to fit your operation. Many farmers do a combination of several markets: they may develop a CSA and sell at farmers' markets and through a broker. Cities and heavily populated areas definitely raise the pricing structure, but travel costs and time spent at the market can eat up the higher returns. It is all a process of finding the best fit for the individual farm.

If there is a large ethnic population in your area, find out about the kinds of food they would like to see at the market. Everyone hungers for something from home. Perhaps a particular kind of melon or cucumber is difficult to find. Grow it and then let it be known that you do.

BRANDING

Let's talk about branding. A farm name is a good idea. If it is the farmer's last name followed by *Farm* (for example, Hurst Farm), that's okay. Or you can get creative: Sunny Dale Farm, Lotta Wata Acres, etc. This name of your farm will become your brand. Think about how much time, effort, and capital major food producers put into naming their product and carrying their brand throughout their marketing strategy. The small farmer can do the same on a smaller scale. Business cards are a reasonable form of advertisement. If there is one piece of advertising that can be obtained for a very low cost, this is it. The business card also lends an air of professionalism, and it establishes a business identity. A sign stating your farm name at your booth or roadside stand

is also important. Let people know who they are doing business with. This will help develop a loyal market. If labels or bags are to be purchased, investigate the costs associated with personalization. Sometimes a label makes it home when nothing else does. What if the customer loves the product and wants more? Make it easy for them to come back!

An attractive logo would also be a good addition to your overall marketing strategy. Anything you can do to implant an image of your farm in your customers' minds will help them find their way back to you. Choose a logo that suites your image. Make sure it is not too cluttered. Use the logo right along with your farm name every chance you get. People will come to recognize your products instantly simply by your logo.

1 "Fruit and Vegetable Prices," US Department of Agriculture Economic Research Service. Last updated September 18, 2013. http://ers.usda.gov/data-products/fruit-and-vegetable-prices.

2 "Commodity Areas," US Department of Agriculture Agricultural Marketing Service. Last updated April 7, 2014. www.ams.usda.gov/AMSv1.0/ams.fetchTemplateData.do?template=TemplateA&navID=FruitandVegrightNavlinkHelp&rightNav1=FruitandVegrightNavlinkHelp&topNav=Help&leftNav=&page=FruitandVegetable&resultType=&acct=fv.

3 "Meat Price Spreads," US Department of Agriculture Economic Research Service. Last updated May 15, 2014. http://ers.usda.gov/data-products/meat-price-spreads.

Goatsbeard Farm

Hand Ladled Rounds
Plain, Herb d Provence
and Black Pepper   6⁰⁰/ea

Marinated Round ——— 7⁵⁰/ea

Fresh Tubs - Plain,
Garlic Herb or Chipotle   6⁰⁰/ea

Feta ——→   6⁰²/ea

CHAPTER 3

IS THIS YOUR MARKET?

Jenn Muno of Goatsbeard Farm, in Harrisburg, Missouri, has been selling at the Columbia Farmers' Market for nearly twenty years. Jenn and her husband Ken make artisan goat milk cheese. Note their signs, which clearly list products and prices.

Chances are there are established farmers' markets in your area. What do you look for when deciding upon your proposed sales venue?

Look for a market that has good traffic, an organized way of doing business, and hours that can fit your current mode of operation. Consider the type of market you would like to shop in. (See Worksheet 5, Appendix 1, for a checklist of what to look for in a market.) Talk to vendors within the market and take in the overall tone. There can be an unbelievable amount of drama at some markets. Unless that is your style, look for one that is low-key, where the majority of vendors seem satisfied, and you note a feeling of unity. Saturday markets are very popular. However, neighborhood midweek markets also draw a crowd. Check with your state farmers' market association to learn about the available markets in your area (see Appendix 4 for a list of associations by state).

LEGAL ASPECTS OF FARMERS' MARKETS

Many markets operate with a board of directors, a market master, and an assistant market master. Some form a not-for-profit cooperation to take advantage of tax credits, apply for grants, or make contributions tax deductible. The market master and assistant are sometimes paid positions, otherwise these positions are held by volunteers. It is helpful to have a board involved in the decision-making processes. If all the responsibility were to fall to the market master, it could be an overwhelming role to fill. A board of directors is also helpful with overall planning and follow-through.

There will typically be an agreement to sign—between you as a vendor and the market as a whole. (See Worksheet 8, Appendix 1, for copies of actual agreements.) Some

are quite lengthy, covering practically everything under the sun, and some are very brief statements outlining the rules and regulations of the market.

INSURANCE

Many markets do require product liability insurance. Generally, the market will carry a policy to cover accidents and liability of the market. However, increasingly markets are requiring liability on the products to be sold. This is spelled out in the contract. Markets are subject to federal, state, county, and municipality laws and regulations. The county or city health inspector has the final rule over what may and may not be sold at a market. The county health inspector is your first contact within the system.

BAKED GOODS

Some markets allow baked goods to be sold that were made in a home-based kitchen; others require a certified kitchen. In the end, it is up to the county or city codes. These codes vary greatly, so personal investigation of the codes in your area is warranted. If there are questions, check with the local health inspector. Meat, poultry, eggs, and dairy products are typically governed by a division of the Department of Agriculture, the Department of Health and Senior Services, or both. These products typically require state licensure or inspected facilities meet production codes or both. The hierarchy of regulators is as follows:

With the high demand for certified kitchen space, many communities are coming together to have a kitchen certified and then renting the space by the hour. This can be an economical way to prepare food, according to law, without all the upfront investment. If there is not a kitchen available in your area, consider contacting a local church to find out about partnering. There are many kitchens that sit unused for the majority of the time. Often, doing a few upgrades is all that is necessary to become certified.

Markets may be designated as producers only. Only items grown or produced by a farmer may be offered at this type of market. Other markets may open up the arena to include vendors who bring in produce from other growers. Make sure your program fits the requirements of the market(s) you apply to.

CRAFTS

Handmade craft items can add to the overall flavor of a market. However, it is recommended that only a percentage of the booth space be made available to high-quality craftspeople. If these applications are not monitored, a market can quickly become a craft fair rather than a farmers' market. A few well-placed booths offering things, such as baskets, handmade shopping bags, aprons, and other craft items can definitely add variety and appeal to a marketplace as a whole. Some other types of handmade crafts might be soaps; candles; woodworking; metalwork or hand-forged items; glasswork; pottery; wild-crafted wreaths; woven rugs; knitted hats, scarves, and socks; and a myriad of other quality crafts. Ask craftspeople to submit samples of their work for jurying. Poll the board members and make selections according to majority rule.

YOUR BOOTH

After you decide on the type of market you want, take another walk through a market and look at the various presentations vendors use to draw the eye to their goods. Well-presented food and crafts in abundance certainly make a good first impression. Some growers focus on one or two items, others have a plethora of produce. Simple changes in presentation can make a huge difference in the overall appearance of a booth. One local grower stated, "Carrots and beets always sell. Leave the tops and greens on. For many, that makes a statement these are fresh and homegrown."[1]

The booth is your storefront. Walk through the produce department at local grocers to get ideas for your booth. Big-box stores spend a lot of time and money

Hierarchy of Regulations

Federal

State (must be as stringent as federal but can be more so)

County and/or city (must be as stringent as state and federal but can be more so)

Market Money is Community Money

Local restaurants

Farm supply store

Farmers' market $$$

Car and truck repair

Hardware store

assessing the optimum methods of display for every product. Mimic some of their display methods. Tables, a canopy, brightly colored tablecloths draped to the ground, your sign with the farm name, business cards, perhaps some baskets to display the produce—all will make a statement that this mini store is open for business.

Choose displays with the thought of portability in mind. You do not want to move in heavy fixtures each week. Displays should be light enough to move but heavy enough not to blow over. Check yard sales or thrift shops for lightweight cabinets, wire shelving, or wooden racks to help display your wares to the best advantage. Remember, there will be children at the market, so display accordingly. Kids can help sell merchandise, but they can also inadvertently damage fragile goods. Keep this in mind when setting up your booth. For a checklist of items to bring on market day, see Worksheet 6, Appendix 1.

KEEP YOUR CUSTOMER IN MIND

As you are building your market business, ask yourself about the type of customer you are attempting to draw into your booth. To state the obvious, you are looking for the one who has money to spend and wants to buy your product. What can you do to bring that customer to you? Where do those people shop besides the farmers' market? What do those shops look like? Are they rustic, country, or is green the primary focus? Learn to read your customers and work to appeal to those who come back week after week. You will learn about the product base that appeals to them. Ask them questions about other items they would like to see in your booth. Give them a sample of something that they might be interested in. Let them know you were thinking "especially of them" when you packaged that sample. Personalization goes a long way. Customers have hundreds if not thousands of places where they can choose to spend their dollars, and they have chosen to spend them with you. The attitude of gratitude should prevail.

MERCHANDISING YOUR CRAFT BOOTH

Crafts add zip and personality to a market! Make sure your booth shows your wares to their best advantage. If you plan to have a canopy, check with the market master to see if there is a specific canopy the market uses to create a uniform display. If not, look for a serviceable and sturdy canopy that one person can put up (unless you have partners). Secure the canopy corners with concrete blocks, gallon water jugs, or other heavy objects. Canopies tend to get caught up in the wind! Make sure not to create a tripping hazard. Bring your tables and brightly colored tablecloths. If you have small objects, consider risers or upturned wooden crates to add height and variance to the display. Include price tags or signs on your work. A laminator is a good investment. Make printed signs and laminate them for use week after week. Often, people are reluctant to ask about pricing.

Hand out your business card. Talk to the people who enter your booth and make them feel at home. Create a photo book with samples of your work. Decide beforehand if you want to take special orders. If you are making a product that lends itself to it, such as soap, hand out samples. Give people ideas of how to use your products. If you make baskets, bring along props, such as a bottle of wine or a loaf of bread. If you make pottery, bring an attractive array of fruits or vegetables to add color and dimension. Remember to bring bags, tissue or wrapping paper, plenty of change, and your receipt book. Some artists use a high director's chair to sit at eye level with their customers. This is a good alternative to sitting at a low vantage point or standing during the entire market.

Strongly consider accepting credit cards through a smartphone system. If you have items priced above the amount a customer would normally bring to a farmers' market, it is important to capture that impulse sale. Do scale your prices, overall, to fit your venue. Bring two or three more expensive pieces, and make sure they are well displayed to prevent being swept up in sudden wind gusts. Museum putty, a putty made for earthquake zones, can be used to anchor items that are at risk of being blown away and breaking.

DISPLAYING YOUR GOODS

Abundance sells. Pile it high and watch it fly is an old adage at the farmers' market. Continue to fill baskets and crates throughout the market day to create the illusion of fullness. If you run short of produce, fill in the blank spots with paper and lay the vegetables on top. Keep downsizing containers as you sell out of particular items. No one is excited about goods that are picked over, so the key is to keep things looking fresh and appealing. Some vendors bring a spray bottle and mist their vegetables to give them that just-picked appearance.

If you have packaged goods for sale, consider the final appearance of the product. Is the product visible within the bag? Does it look fresh and attractive? Consider tying the bag with a piece of raffia just for that extra touch. At a recent farmers' meeting, one vendor commented that if she packaged brownies and added a bow, she could significantly increase the price of the brownies. A small investment in ribbons, twine, or raffia can have an impact on the bottom line. It is always those extra efforts people notice and appreciate. This is part of what makes your market "store" unique. Customers will remember those little extras and come back time and time again. This is one of the many things that differentiates a market from a big-box store. Adding this kind of final touch can also reinforce your branding. "Let's go to the booth where the lady ties red ribbons around her brownies." A lasting impression made and sold.

BEHIND THE BOOTH

Make sure to have change, a receipt book (if needed), a calculator, bags, and whatever is needed to conduct transactions. Vendors frequently accept credit or debit cards. You can download apps, such as Square (www.square.com), to your smartphone that for a small fee per transaction allow you to swipe credit cards as payment. Taking credit cards can boost overall sales.

High tunnels are basically unheated greenhouses constructed of metal hoops and supported by cross members and wooden framing on the bottom and side walls. They are covered by one or two layers of plastic and are heated by the sun. It is simply amazing that a layer of plastic can boost temperatures enough to extend the growing season by several weeks. Conversely, in areas of extremely high temperatures, the same structure can be used with a shade cloth to protect plants from the heat.

High tunnels provide protection from inclement conditions, making them the perfect environment for cut flowers, tomatoes, and other high-dollar crops. Imagine walking into your tunnel on a February morning in the Midwest and taking in a breath of fresh, sweet earth. To see something growing at that time of year is a gift in itself.

Consider applying to accept supplemental nutrition assistance program (SNAP) benefits. This will require a terminal to process the electronic benefits transfer (EBT) card. The US Department of Agriculture (www.fns.usda.gov/snap/retailers-0) administers this application process. By accepting SNAP at your market, you are offering fresh food to people from all walks of life and economic situations. This is a benefit to all.

The demeanor of the vendor also plays a vital role in overall approachability and, thus, overall sales. Think of the last time you entered a store where the employees were lackluster, sitting down behind the counter and giving off a vibe that your business really didn't matter to them. What was your impression? Will you go back there again soon? The beauty of the farmers' market atmosphere is the opportunity to connect with the customer. This is the time to be sincere and engaging and to promote yourself, your farm, and your products. Some farmers bring photo albums for customers to flip through, tablets with slide show presentations, or other means of educating their customers about who they are and what they do. Customers do want to know who they are buying from and how the product was raised. This promotes sales, loyalty, and goodwill.

Offer recipes. A few printed cards will go a long way in assisting your customers in preparing their purchase. If it is the height of tomato season, bring a recipe for homemade ketchup or chili sauce. This is especially helpful if you are growing some of the lesser-known vegetables. People enjoy trying something new, if they know what to do with it. Help them learn.

Chatting with and moving customers in and out of the booth in a timely manner is an art in itself. It is helpful to have a partner in the booth to handle actual sales transactions while the other member of the team serves as the "front man." There is a bit of theatrics involved in selling at a market. Customers go there because they want something more. When you hear the word *farmer* you think friendly, wholesome, real. Being memorable will pay off. Attempt to learn your regular customers' names. Have a sign-up sheet so you can build a mailing list. Offer a raffle item each week and collect the names of people who sign up. It won't be long before you have a massive mailing list. Especially collect emails to eliminate costly postal delivery. If you add to your list each week, you'll soon be able to send out messages to thousands of contacts with one keystroke. This keeps people connected and involved in your business. Customers will want you to succeed.

DURING THE WINTER

Many markets are turning to year-round events. For those living in areas where winter is a factor, this is a fairly new development. Some markets move indoors to a warm and welcoming spot; others brave the cold and loyal customers bundle up and keep going to the market.

continued on page 46

FARMER SPOTLIGHT

Carl and Robin Saunders

Carl Saunders of Yellow Dog Farms with his primary crop of salad greens.

Carl and Robin Saunders of Yellow Dog Farm utilize high tunnels to the fullest. Carl started farming as a retirement project. He has never worked harder in his life! Coming from a retail background, Carl and his wife, Robin, wrote a business plan for their farm in 2006, and they have never looked back. They started farming because of their interest in local foods and food production. "We bought 30 acres and we wondered what we could do." At the heart of their farm is a series of high tunnels that expands their season by thirty days in the spring and thirty days in the fall. The couple produces an amazing amount of food on a relatively small piece of land. And, yes, there is a yellow dog. Hank, a yellow Lab, serves as mascot.

"Our goals were to remain small enough to be able to work the entire operation as a couple. I am always looking at what I can do to level the workload. The high tunnels assist us in equalizing our labor," states Saunders.

The Saunders begin their season in late February by turning on the heat in the greenhouse to start their seedlings. "We harvest our first salad greens in the high tunnels in late March or early April. We continuously produce these greens until mid-June. Then we replant in August for harvest from October through mid-December. Using high tunnels we are not forced to wait on the weather. With sequential plantings we can offer a continuous supply to our buyers." The Saunders sales are primarily through the Lake Saint Louis Farmers and Artists Market (www.lakestlouisfarmersand-artistsmarket.com) and through CSAs.

In addition to the indoor growing space, Saunders farms five acres of vegetable crops. Recently he planted 9,000 pieces of garlic. "My customers are enthralled with garlic. They just can't seem to get enough of it," Saunders states.

Saunders is the driving force behind the Lake Saint Louis market, serving as market master and overall cheerleader. The area is affluent, and it has embraced the market. "On a typical Saturday morning, we have about 3,000 to 4,000 people attend the market. Traffic is great." Carl attributes this to a prime location, community support, and a varied number of goods for sale. "Vendors bring their best," he remarks.

The market has progressed at a remarkable pace, due in part to the location in the Meadows outdoor mall parking lot. "The Meadows has been great to work with us." Last season the market and the Meadows sent out a direct mailing to 50,000 homes in the area, noting the market times and featured products. "This brought in many new customers who lived less than a half mile away and were not aware of the market. This effort was well worth the dollars spent," remarks Saunders.

The market seems to have hit upon a good balance of farmers and artists. Glassware, pottery, wooden crafted items, baskets, handmade soaps, baskets, handmade socks, and other items are included in the mix. It's truly a one-stop shop for gifts, produce, honey, eggs, meat, and handmade bakery items. The market received a 501(c) (4) status and operates under a board of directors as a not-for-profit entity. "Our biggest problem right now is that we need more farmers. We can always use more farmers!" remarks Saunders. "The demand is always there."

Carl states, "It is a relatively new skill for farmers to represent themselves at a market. People are looking for that connection between their farmer and their food. Being a vendor at a market requires an uncommon skill set, first that of grower/producer, then that of marketer, and finally, as a salesperson. It is something that develops with time." The Lake Saint Louis Farmers and Artists Market runs from April through October, Saturday mornings from 8:00 a.m. to 12:00 p.m.

continued from page 43

Season extension through high tunnel or greenhouse production allows the growing season to last well past the last frost date and begin well before spring production is normally feasible. These winter markets keep customers in the habit of shopping the farmers' markets for their goods.

It cannot be denied that high tunnels have opened up a whole new world (and extra season) for farmers. Customers seek local products year-round and, of course, as demand goes up and production is limited, the price for goods increases exponentially. If you have the first tomatoes, you will get the highest price. If you have the last tomatoes, you will get the highest price.

Some markets have special holiday events with markets just before Thanksgiving. Think of squash, pumpkins, Brussels sprouts, and sweet potatoes rolling into harvest baskets. And what better way to celebrate the holiday season than with baskets full of local goods. What a unique and lovely gift, sure to be appreciated.

Winter markets are a bit harder to handle than warm season markets. Some outdoor markets invest in tents and heaters to keep their vendors and customers at least somewhat warm. An indoor space takes a bit more finesse requiring displays, lighting, and space. However, this type of market is worthwhile for several reasons, including the following:

Winter markets keep customers in the habit of coming to the market.

For those with late season production, the winter market provides a place to sell produce.

There are opportunities during the holidays to sell items, such as hams, turkeys, geese, and ducks, gift baskets, and handcrafted items. Candy is also a winter boon.

COMMUNITY BENEFITS OF FARMERS' MARKETS

The farmers' market was once a hub of activity, and this environment is returning with the advent of music, art fairs, coffee booths, and ready-to-eat food. This creates an environment where people want to be. Statistically, the longer people shop, the longer they have to buy. One popular market in the city of St. Louis offers a wine garden with entertainment every Thursday evening. This makes a great venue for selling local wines, locally made cheese, breads, desserts, and more.

Farmers' markets are good for communities. A vibrant market will attract a multi-generational crowd: young mothers buying fresh produce to make their own baby food, families with young children, seniors out for a Saturday morning adventure. Markets promote the sense of neighborliness, of localism, and of old-fashioned commerce. Customers feel good about supporting local farmers. It is a win-win situation for all involved.

A large percentage of dollars spent within markets stay within the community. The producer sells goods to the customer and collects the funds. Those funds, in turn, go back to the local farm supply store, the area garage, the hardware store, etc.

YOUR BOOTH

Fees for booths vary greatly. Reports show everything from free market spaces (intended to establish a market) to major investments for a market stall. These fees should be relative to the size of the market, the overall potential to bring in customers, and the ability to create an income for the participating farmers.

If there is more than one market in your area, talk to vendors already participating in these established markets and find out if they have been there for several years and how their fee-to-income ratio works out. Consider your projected inventory and whether you can meet the demands of a high volume market. Remember, full and well-stocked tables draw the customer in.

Make it easy for your customers to find you. Try to secure the same booth space each week or at least attempt to be in the same vicinity. Use a colorful flag with your logo or other means of instant identification for your booth. Make an attempt to learn your customers' names and use them. If you have various family or staff members manning the booth, let them know your mode of operation and how you want your business to be conducted. Remember, your booth is your storefront. Empower your workers to make decisions so transactions can be handled in a timely manner.

Sampling may or may not be allowed per the individual market's stand on this issue. If sampling is allowed, then follow the market guidelines to ensure customer safety. Sampling is proven to boost sales of the featured product, so be prepared to sell an abundance of the item you are offering for sample. Recipes to go along with the samples are always well received.

Regional products are to be celebrated. What is special from your area? Who can resist Vermont maple syrup, Tupelo honey, Florida and California citrus? Because you live with the foods of your region each day, they can be taken for granted. Local, local, local. It's big business! You should know your product backward and forward. Be able to answer all questions about how the item was raised, the folklore (if any) surrounding the item, and offer serving suggestions and recipes to support the sale.

Customers may become irritated when items go on sale at the end of the day.

Establish your price and stick with it, from the beginning of the market to the end. Establish per-pound, per–piece, or per-box prices. If goods are sold by the pound, then an approved and inspected scale is required. Check with your state's division of weights and measures to find out what type of scales are eligible for certification and where to go for that service.

Often markets partner with food pantries, and at the end of the market day, unsold goods are donated to a pantry. This is a win-win situation: the community is served with high-quality food and the not-for-profit vendor gets a tax deduction for the donation.

If there is a downside to this type of marketing venue, it is the amount of time invested. Plan on spending at least one full day of picking, bundling, and prep work. Then there is the drive to the marketplace, the setup, the sales time, tear-down, and travel home This type of venue can simply be exhausting. Some people thrive on it, but others don't enjoy it all.

Weather is a factor; high winds, rain, and other inclement conditions can wreck canopies, displays, and attitudes! If you plan to use a canopy, figure out a means to anchor it down. Expect downpours, wind, and other inconveniences. Seasoned farmers' market vendors know these are simply facts of life. Customer safety, as well as your own well-being, is vitally important. Anchor things down as much as possible, and then keep an eye on the weather throughout your day.

Pick your market carefully. If the farmers' market is not your thing, perhaps a CSA would be a better option.

1 A note about processing. If vegetables or fruit are sold sliced, chopped, or altered in any way, then they are considered to be processed. Simply present the vegetables and fruit whole if the goal is to sell fresh produce.

CHAPTER 4

SUCCESSFUL
FARMERS' MARKETS

Color and variety sells. Note the unusual vegetables such as kohlrabi. Offer recipes and serving suggestions.

Local goods, satsumas, lemons, grapefruit, and other southern-grown fruits are available at the French Market in New Orleans.

When figuring out a formula for success, it is important to look at other markets that have been around for years. The four markets featured in this chapter are in metropolitan areas. All four markets were chosen for their unique atmosphere and approach to marketing. The French Market in New Orleans, Soulard Farmers' Market in St. Louis, and the Portland Saturday Market in Portland are held in major cities. Burlington Farmers' Market in Vermont is the closest of the four to a traditional farmers' market.

THE FRENCH MARKET IN NEW ORLEANS, LOUISIANA

If you have never experienced a cup of chicory coffee in the heart of the French Quarter of New Orleans, then add this to your bucket list. For more than two hundred years, this market has been a bustling site of commerce. Arguably one of the most historical market sites in the US, the French Market's history spans centuries, not decades. In the late 1700s, "Spanish authorities ordered the construction of an enclosed market to prevent the continued exposure of food to the elements at the established outdoor riverfront market."[1] This market has endured nature's full force, withstanding hurricanes, fires, and floods over the years. The French Market was one of the first businesses to come back to life after Hurricane Katrina. It stands as a monument to the resilience of the people of the city.

Perhaps what makes New Orleans so special is the blend of people. Since it was founded, the city has served as a true melding place, a melting pot, drawing people in from far and wide. The Creole, Acadian, French, African American, Spanish, and other nationalities each add their own special flavor to the food and the overall flavor of the area. The various influences are alive and well in music, art, culture, and food in this special place.

At one time, the French Market was the central trading post for New Orleans. Sitting on a port of the Mississippi, it was the hub of commerce, seeing goods and people pass from all over the world. Practically everything imaginable was sold at the market at one time. Wild birds, vegetables, fruits, dry goods, wild game, and exotic imports were offered in the market stalls. This market went through various stages of development, progressing with the times.

In the 1930s, a Works Progress Administration (WPA) project added the now famous Bazaar Market with areas specifically for fruit and seafood vendors. New and updated sanitation and refrigeration was added at this time. In 1975, another major renovation took place, and the focus of the market changed from wholesale food distribution to retail shops and restaurants. The farmers' market remained and soon the flea market came into being. In 1991, the market went through another renovation in preparation of the 200th birthday market celebration. The aftermath of Katrina brought about another time of refreshing and regrouping for the market.

Not surprisingly, coffee always had a role in the French Market. Perhaps it was to assist those who had imbibed a little too heavily at some of the local establishments the night before, but the special blend of coffee and chicory is a New Orleans' favorite.

Café Du Monde still holds court in the French Market. Established in 1862, this landmark is a must-see for tourists and locals in the French Quarter. Café Du Monde is a marketer extraordinaire with its brand evident throughout the Quarter and the city proper. Take-home coffee and beignet mix can be bought onsite. These products are also available throughout selected retailers in the United States.

Nothing can take the place of enjoying a café au lait in the heart of the French Quarter. It is impossible to talk about New Orleans and not discuss food in general. Muffulettas, poor boys, crawfish étouffée, and other specialties can all be enjoyed at the French Market. And then there are pralines. These are best when eaten in a horse-drawn carriage on a tour of the area. Don't miss the silver-painted mimes.

Locally grown goods are still available, and you are likely to find alligator jerky and other regional favorites. The overall tone of the market is one of a happy place that brings people back again and again. Local artists love this market, and many come back day after day to display and sell their wares. A unique blend of imported goods, locally produced art, music, and regional foods attract tourists from all over the world. Once you have visited this city and, in particular, the French Market, you will want to return time and time again.

BURLINGTON FARMERS' MARKET IN BURLINGTON, VERMONT

The Burlington Farmers' Market in Burlington, Vermont, is a warm and welcoming venue. Vermonters are proud of their locally produced goods and happy to share their bounty with an enthusiastic customer base. Formed in 1980, this market features a touch of a country fair feeling combined with an upscale city market. A plethora of goods awaits the Saturday morning shopper, everything from artisan cheeses to fine chocolates, hot sauces, handmade goods, and more.

Located in the heart of Burlington, the city park offers the perfect hometown venue during the warmer months, and the Memorial Auditorium houses the market during the winter. The winter market is held every other Saturday from November to April. Snap and debit cards are accepted. A market manager, two assistant market managers, and a market assistant oversee the week-to-week operations. A steering committee keeps the market on track. Farmers, artisans, and community representatives are represented on the board to provide a well-rounded voice.

Vendors come from all over the state. With over eighty booths, there is something

A plethora of fresh vegetables awaits shoppers at the Burlington Farmers' Market in Vermont. Note the artful display—a sure way to attract customers.

for everyone. Of course, locally produced food is the main feature of the market, and there is plenty of maple syrup, apple cider donuts, and other food only produced in Vermont. This market is also unique in that customers can purchase milk, meat, bread, eggs, vegetables, and seasonal fruits. It is possible to fill a whole grocery order with one stop. (This is a goal for many markets.)

This market provides the type of atmosphere that makes people want to linger, see old friends, and get to know the farmers, producers, and vendors. Musicians, songwriters, and performers are featured each week.

New and existing markets can learn from the success of this market. Weekly entertainment, food, festivals, and fun make for a well-attended event.

Burlington is the largest city in Vermont and the site of the county seat. A college town, Burlington is progressive and has been at the heart of the local foods movement since the 1980s when the farmer's market began.

PORTLAND SATURDAY MARKET IN PORTLAND, OREGON

The PSM is unique in that it is all about the art. This market is nationally recognized as the largest continuously operating open-air arts and crafts market in the United States. The artwork is complemented by live music and exotic foods available every Saturday from March through December. The 2014 season marks the 41st anniversary of this weekly event.

Operating as a not-for-profit, the PSM is home to over three hundred artisans. Developed as a place where artists could come and set up their "storefronts," many run their small businesses each and every Saturday. Countless artists make their primary living from income generated through their weekend sales. Every item offered at the market is handmade by the artists.

In 2006, in a redevelopment project of the Old Town district, the PSM established a permanent home in Waterfront Park. PSM now generates sales of over $10 million in gross sales on an annual basis, attracting 1 million visitors annually. The PSM board of directors continues to be made up of a majority of market members, putting market governance in the hands of its members. Seven full-time and ten part-time staff members administer the operations and various programs of the market. Started with grant assistance from the Metropolitan Arts Commission, the Portland Saturday Market has earned a national reputation for service to artists.

It is nearly impossible to mention all the wonderful works to be found at the PSM: stained glass, handwoven items, sculptures, photography, jewelry, pottery, woodwork, soaps, and so much more. The market is juried to ensure that the highest-quality product for consumers is available.

Fresh vegetables are just the tip of the iceberg at Soulard Market, which is open Wednesday through Sunday all year.

Portland is Oregon's most populated city, and it continues to grow at a pace above the national average. The PSM draws vendors and customers from all over the Pacific Northwest.

SOULARD MARKET IN ST. LOUIS, MISSOURI

This market is mostly under cover yet still has many open-air sections. There are permanent merchants with stores in the Grand Hall, including butcher shops, florists, a spice shop, bakeries, and hot-food marts. The blend of vegetable marketers, fish mongers, cheese makers, soap makers, and

OPPOSITE: Freshly picked lavender awaits the customers at the Portland Saturday Market. A great variety of art, handmade items, and food is available every Saturday and Sunday at this outstanding market.

other market participants add to the color of the vibrant space. This market welcomes vendors with purchased goods as well as those who grow their own or make their own products. With over eighty vendors hawking their wares, this market is open year-round Wednesday through Saturday. Many market shoppers enjoy a Bloody Mary while conducting their Saturday morning rounds.

The market is located in the historic Soulard area of St. Louis, which is home to the second largest Mardi Gras celebration in the country, outranked only by the Mardi Gras in New Orleans. The shops and restaurants of the area are sure to provide something for everyone. Soulard is as much an icon in this city as the famous Gateway Arch.

1 John Magill, "French Market Celebrates 200th Anniversary," Preservation in Print 18. no. 4 (May 1991): 2.

Seafood & Cajun Market

Local Farmed Gator

ONLY
SOLD
HERE !!

CHAPTER 5

STARTING A FARMERS' MARKET

If you can get four to eight people together, each with fresh, local produce to sell, you can start a farmers' market. Even if you don't have fresh gator meat, it can still be a smashing success.

If there is not a viable market in your area, consider starting one. The best way to begin is to contact friends and neighbors who grow, bake, or make quality handmade items. Plan a meeting and see what kind of community interest there is. If there are enough (four to eight people can start a market), then you can move forward. See Worksheet 7, Appendix 1, for a to-do list when starting up a market.

FIRST STEPS

There will have to be some kind of governing body—either a board or one person as the market master—to give structure, answer questions, talk to the press: in other words, present a united front. It is also advisable to have the vendors sign an agreement stating what can and cannot be sold, hours of participation, cost of and payment for the booth fee, etc. (See Worksheet 8, Appendix 1.) Will you require product liability insurance? Depending upon your location, the landowner may require liability insurance. It is definitely advisable to have liability insurance on the site.

Start your planning sessions in the fall before your spring opening. It takes a surprising amount of time and preparation to get started. Established and successful markets prove that this forecasting is time well spent. You do not want to start the first day without a pretty good idea of how things are going to run. First impressions are lasting to your customers. After a few weeks, it will become natural, although there are always unanticipated events when you are dealing with the public.

If you are starting a market, decide how you want it to look. What kinds of displays will be expected? This artisan is taking advantage of all available sales space in an urban parking lot.

Competition for farmers can be a factor. In almost every market, there is a reported shortage of producers. What can you offer a farmer to set up a booth at your market? Good location, advertising, perhaps assistance in setting up or taking down displays? Of course, no one can guarantee sales and part of that rests with the producer. Simply work to provide a hospitable environment, a friendly well-run market, and overall support.

Rural and city markets differ namely in the pricing structure of goods sold. It seems the farther away customers live from a farm, the more they are willing to pay for the products. Perhaps this is only fair when considering travel time, mileage, gas, etc. There is a price to pay for participating in high-end markets, both in expenses and fees. Make sure to consider all costs before traveling too far away from home. For some, a trip to the city is well worth the effort.

Decide upon the days and hours of operations, and such concerns as weather policies. What type of displays will be expected? Will vendors provide their own amenities, such as tables and canopies? Will a specific type of canopy be required or can vendors use what they have (are you going for a uniform appearance)? Will this be a producer-only market or can vendors bring in produce bought from other sources such as brokers? What do you consider local—20 miles, 50 miles? All of these questions should be considered at your first meeting. Appoint a secretary to take notes, then distribute those notes so everyone is on the same page. What percentage of booths will be designated for crafts? Will you supply electricity? Decide upon a list of rules (see Appendix for examples).

Alert the local health inspector of a coming market. Work with the inspector from the beginning to foster a good relationship. This is key to getting along with the powers that be from day one. Have the health inspector join the potential market vendors at the second meeting and discuss rules and regulations so everyone is on the same page. Talk about restrooms, hand-washing stations, and the requirements for meeting the city, county, state, and any other pertinent codes.

If your city gets on board, so much the better. Often the area chamber of commerce or other city government entity is glad to assist with some resources and expertise. Ask for help with such start-up needs as procurement of liability insurance and a banner to announce the market. Markets are beneficial to the community, and the more buy-in you have, the better. Civic groups often volunteer to lend support.

RAISING MONEY AND ADVERTISING

Every organization needs money: incidentals and rental fees for facilities and equipment, to name just a few expenses. Farmers' markets operate by collecting fees from vendors, but this is often at a shortfall of actual operating expenses. Consider fundraisers. Sell T-shirts, shopping bags, raffles for market baskets, and other inventive ideas to raise funds. Ask vendors to participate by offering a special farm visit for raffle, perhaps a farm-to-table dinner or another event. Ask other organizations to assist with fundraisers, too. After all, the farmers' market is a benefit to all.

Contact the local newspaper and radio stations to let them know about the new market. Engage in shameless market promotion! Start a Facebook page for the market. Do what you can to promote your market, even if it is on a shoestring budget at first. Designate a portion of the vendor fees for advertising. Buy a banner and signs as soon as possible. Consider a publicity committee and, find someone with some talent for writing. Create a market newsletter to keep in touch with vendors and customers.

If you have an advertising budget, use it to post small ads in the local newspapers to inform potential customers of your market. Utilize opportunities for free publicity. This can come about in a variety of ways. Host contests and ask the press to come to photograph the winners (best salsa, weirdest squash, biggest pumpkin). Invite local dignitaries for the opening of the market ribbon-cutting ceremony, and inform the press. Later, host a customer appreciation day and give awards to those who have supported the market throughout the season.

Markets are newsworthy events so be sure to invite the press.

Often Boy Scouts, Girl Scouts, and 4-H clubs are looking for volunteer opportunities. Ask these groups to come in and help tear down after the market. They can also come in and do demonstrations of crafts, gardening, and other service projects. Try to include other organizations in the market to build strength, traffic, and community commitment. Assign jobs and delegate responsibilities.

THE MARKET MASTER
Find the right person for the market master job, whether this is a paid or an unpaid position. A dynamic person—someone who

HOW TO WRITE A 'NEWSLETTER'

- Find a newsletter template online.
- Start with a one-page (front and back) format.
- Create a header with your farm name, logo, and contact information.
- Write the newsletter with the current season in mind. Include details about your farm but don't get too personal. Write about your crops, your animals, and other pertinent details.
- Include a calendar and make note of the farmers' markets you will be attending.
- Include a recipe for a seasonal dish.
- Note any special events at your farm.
- Note special days such as National Tulip Day (or make up days to celebrate your specialty crop).
- Include plenty of pictures.
- Publish your newsletter on paper to hand out at the market, and archive them on your Facebook page.
- Send the newsletter to your mailing list as well.

is good with people and firm but fair—is the perfect market master.

The market master rules over all on market day. The reason the market master must have ultimate authority is because market day is not a day for debates. If there are issues, they should be addressed after the market, not during. Once again, on market day the market master rules: no ifs, ands, or buts. If the market master must be absent, then an assistant should take his or her place, and this assistant will have the same authority. Conflicts at the market are very bad for business. Hopefully the work done beforehand to organize the group and promote cohesion will keep the group working together in a successful, cooperative manner.

KEEP YOUR CUSTOMERS HAPPY

Make sure to adhere to rules. If the market is to open at 7:00 a.m., make sure it does. Vendors should not be allowed to sell before the market opens. This prevents customers from coming in early and buying up the "good stuff." If the market is to remain open until 1:00 p.m., make sure it does. There is nothing more disappointing than making a last minute run, arriving at 12:50 only to find everything closed down. Some vendors say they do their best sales at the close of the market day.

Add a welcome/information booth to your market. Offer coffee and fresh pastries for sale (as a fundraiser) or invite a vendor

At Soulard Market, these beautiful piles of oranges and lemons are sure to grab shoppers' attention. Remember: Abundance sells.

HOW TO SET UP AN ATTRACTIVE MARKET BOOTH

- Put up an attractive canopy. If yours has been through several seasons and looks shopworn, then purchase a new one for the upcoming season. See if the market has specific requirements or recommendations for uniformity.
- Bring tables that don't sag in the middle! Invest in sturdy tables or reinforce existing ones to hold the weight of your vegetables or other goods.
- Use brightly colored tablecloths. Make sure the tablecloth is skirted to cover extra stock or boxes underneath the table.
- Hang a sign or banner with your farm's name.
- Display your wares attractively. If you have fruits or vegetables, remember that abundance sells. Refill as you sell.
- Place signs showing your prices where they are easily seen.
- Wear a T-shirt or apron with your farm's name.
- Offer business cards to your customers.
- Maintain a professional yet friendly attitude.
- Experiment with various table configurations. If you have a standard 10 x 10 booth, then try placing your tables in a straight line, or shaped as an *L*, a *U*, or a *V* and see which best suits your product and overall space.
- V-shaped booths are inviting to people, and they bring customers into the booth before they realize it. After it is set up, view your booth from the outside through the eyes of the customer.
- Eye level displays are attractive; consider vertical real estate expansion with crates and racks.
- Stand up (take sitting breaks in between customers if need be).
- Don't eat at your stand (professionalism sells).
- Bring bags, change, receipts, and other items needed to conduct business.
- Honor your commitment and follow market rules.
- If you are going to miss a week, let your customers know. They will come looking for you. Let your regulars know, then tweet or post on your Facebook page.
- Consider a theme: French country, farmstand, ethnic, etc.
- Clean up.
- Donate excess goods to the food bank. What goes around comes around!
- Smile, no matter what!

to come in and sell these goodies. In hot weather, offer lemonade and iced tea. Vow to make the market the most welcoming, most friendly place there is. If there is a booth that needs a little zip, bring in an extra tablecloth and offer it to the vendor. Offer a fresh flower arrangement, and help that vendor have an attractive and successful market booth. Everyone will benefit from the extra touches. The goal is to optimize the customer and vendor experience, to enhance the overall spirit of the market, and to make yourself and everyone else look good. Raise the bar!

Markets, by nature, draw a crowd. They are generally happy places. Plan events that foster that feeling of light and airy, easy and breezy. Music definitely sets a tone for a positive atmosphere, so schedule local musicians to perform. Pay them, allow them to sell their CDs, or both. Nothing adds to a market like music. Bring in chefs and provide on-the-spot cooking demos. Plant seeds, hire face painters, make a fun environment for kids so when their parents say they want to stop at the market, the kids will be eager to go along. Sell kid-friendly foods and projects. How about kid-sized soap or muffins? And, of course, don't forget berries, cherry tomatoes, and all those other delicious, healthy treats.

MARKET LAYOUT

As you lay out the market, think about traffic flow, where bottlenecks might occur, and how to maximize exposure for all vendors. Certain booths will always be a huge draw, whereas others will fade into the background without proper merchandising and extra effort to offer enhanced visibility. Plan a meeting and offer merchandising advice. Help your vendors be successful. If you plan to have music and activities, designate specific areas for those activities.

The welcome booth should always be front and center—a place for customers to go for directions and helpful suggestions. The person who works the welcome booth should be knowledgeable about the market and have a good handle on the layout and vendors. This is the place to sell those fundraiser items, recruit volunteers, and enhance the sense of community. This is also the place to have a first-aid kit and a lost and found.

If there are several vendors with the same product, space them apart from each other. Be careful that you don't allow too many vendors who sell the same products into the market. Court farmers who have something unique. Mushroom growers, cheese makers, and other purveyors of specialty foods are often in short supply. If you can boast you have these vendors at your market, it will boost traffic and sales. Variety is key. At a recent famers' market gathering, the remark was made that it should be a goal for a farmers' market to satisfy 80% to 90% of customers' overall grocery demands. With a well-stocked, well-rounded market, this goal is indeed attainable.

If your market is a bit lackluster, take a look at the displays. Are they inviting? Well stocked? Fresh and appealing? If not, the market master can make a few suggestions for improvements. These suggestions should be made in a neighborly and diplomatic way, of course. This is when it is good to have one designated person, the market master, to handle business.

SPECIAL EVENTS

Special events are a definite draw. People love competitions. This is a great way to focus on a particular vegetable. Why not

A good example of a double booth. The plants are large and take up quite a bit of space. Doubling the booth space makes an attractive and accessible display.

a zucchini festival with a prize for the biggest zucchini, or a pesto, garlic, tomato, or pumpkin festival? Whatever the season, host a party at the market.

One popular event at an area market is the customer appreciation day. Last year on this special day, hot, sweet corn was served on a stick, dripping in butter. This treat was free to all attending the market that day. Now that's appreciation! As cooler weather approaches, a chili cook-off is always a successful fundraiser. Bring in area teams to compete for the best chili and raise money for the market at the same time. Consider creating a committee within the market group specifically to work on special events and fund-raising. Cajole volunteers with big personalities to step up and lead the march.

MARKETS AS A NOT-FOR-PROFIT

Many markets form as a not-for-profit organization. Exactly what is a not-for-profit?

According to USLegal.com (http://definitions.uslegal.com/n/non-profit-corporation/), "A nonprofit corporation is a corporation formed to carry out a charitable, educational, religious, literary, or scientific purpose. A nonprofit can raise funds by receiving public and private grant money and donations from individuals and companies. Certain federal, state, and local income, property and sales tax exemptions are available to nonprofit corporations. The federal and state governments do not generally tax nonprofit corporations on money they make that is related to their nonprofit purpose because of the benefits they contribute to society. The most common federal tax exemption for nonprofits comes from Section 501(c)(3) of the Internal Revenue Code, which is why nonprofits are sometimes called 501(c)(3) corporations. Tax-exempt nonprofit organizations offer donors an individual deduction for contributions. (Private donors can claim personal federal income tax deductions of up to 50% of their adjusted gross income for donations made to 501 (c)(3) organizations.)"

Under 501(c)(3), a market can accept donations and provide the contributor with a receipt for a tax deductible expense. Another advantage with this type of operating structure is the option to apply for grants. (Most granting agencies or foundations do not offer grants to individuals).

Each state has its own set of requirements for filing for this type of entity, but it is a fairly simple process. Inquire with your secretary of state's office to find the specific requirements. Then request a corporate income tax exemption with the IRS. Employ a tax accountant or attorney to assist you with this process. You will be required to write a mission statement for the proposed organization, form a board of directors, and write bylaws for the group.

HOW TO START A FARMERS' GROUP

Farmers get antsy in the wintertime. A great way to build the strength and community spirit of the market is to create a farmers' group that meets in the off-season. There are always new things to learn about, and this type of group can be part learning opportunity and part social gathering. Poll the farmers from your market about their interests and which programs they would like to attend. Bring in speakers that address those interests. Suggested topics include grafting plants, beekeeping, syrup making, beer making, cheese making, and soap making. Try to keep the presentations to an hour or so, hitting the highlights of the processes. This could lead to a series of classes for those who wish to take their knowledge base further. These kinds of events build camaraderie and allow people to get to know each other better. Behind these efforts a sense of unity grows and people develop friendships that go a long way in keeping the peace at the marketplace. Don't tax the farmers by asking them to get together during their busiest seasons. Keep the classes light yet informative. Cooperate with the group to bring snacks and coffee and to have a nice evening.

This type of structure adds validity to your market and assists with the definition of roles, goals, and operating structure. Keeping minutes of meetings, having a treasurer's report, imposing voting procedures, all help to keep your market a business.

BOARD OF DIRECTORS

Often the farmers' market board of directors and members meet throughout the year. Some meet during off-season months and bring in educational speakers. This assists the group in maintaining continuity and brings opportunities for learning. Community service projects frequently begin with a suggestion from and follow-through by these groups. Although the seasons are busy, farmers are generally a giving group of folks.

It is also prudent to have an end-of-season wrap-up. This meeting should take on a celebratory tone, as everyone has made it through another year! It is also time to congratulate all concerned on a job well done. At this meeting, pass around a suggestion box, then at your first meeting of the new season go over those suggestions to see if there are things that need to be changed.

The first meeting of the new year is about electing new officers, talking over the last season and any needed changes or improvements. It is the time to lay out the market, solicit vendors, and begin market life all over again. Consider running a farmers' education group throughout the off-season. This will assist the group in remaining connected and cohesive throughout the year.

CHAPTER 6

OTHER PLACES TO SELL YOUR PRODUCE

These mixed bouquets were arranged by a bride for her special day. The flowers are from Urban Buds in St. Louis. It's amazing what can happen in a city space with talented growers at hand. Courtesy of Urban Buds

A summer morning on an old Missouri homestead.

COMMUNITY SUPPORTED AGRICULTURE

Community Supported Agriculture, or CSA, was first introduced to the United States in the 1980s. Adopted from European models, CSA marketing is centered upon the purchase of shares of goods from the host farm in advance of the growing season. This method of operation is beneficial to farmers for several reasons:

- Upfront payment covers the seasonal operating costs.
- Farmers will be able to budget their income, having a solid figure to work with.
- Farmers will not have to collect money each week.
- Farmers can charge full retail price with no middleman.
- Farmers will have a good idea of quantities to plant.

Weekly deliveries or pickups can be managed according to time constraints. There is no sitting at a farmers' market for hours each week.

CSAs are advantageous to customers as well. They know that they are buying into a small, locally owned business and that they will be receiving high-quality, fresh foods. Because there is no middleman, they know that prices will be reasonable, and weekly deliveries or pickups allow customers to know when the food is coming and to plan their menus accordingly.

CSAs frequently operate on a weekly basis with a box due to those subscribing each week. Customers buy into the potential bounty and also the potential risks of farming. Farmers can guarantee a certain number of pounds per week or the customer can simply get a share of the available goods. For example, if a farmer has a banner year for tomatoes, customers can expect to see more of those in their CSA boxes. If it is a poor year, the customers will share the loss and have fewer tomatoes. For an example of a CSA agreement between customer and farmer, see Worksheet 9, Appendix 1.

Some CSAs ask for their shareholders to come to the farm and work as a part of their commitment. Other CSAs simply ask for a fee and expect no work in exchange. Be sure to plan on educating those who come to your farm to work. Remind them what kind of clothes and shoes to wear.

CSAs can be stand-alone entities with all the produce coming from one farm, or a farmer may serve as an aggregator and bring in produce from additional farms. A CSA can also be a group effort with several farms participating. Although customers agree to take a risk, if a CSA experiences problems for more than a year, the buy-in from customers can lessen. Customers agree to take the risk, yes, but only to a point. Buying produce from other farms can reduce the pressure to provide a certain number of pounds of food each week. And, yes, there is pressure in this type of arrangement. Several farmers report a great deal of stress in attempting to fulfill orders and CSA shares.

An early spring array of radishes.

In addition to produce, some CSAs offer premiums of freshly baked goods, eggs, even meat and poultry can be arranged as an extended share. Depending on the goods included (and the need for refrigeration), pickup can be at the farm or at a drop-off point in town.

Remember customers do not have a choice in what they receive as their base amount. If it is a good year for cucumbers and a bad one for zucchini, then pickles will be on the menu. This is advantageous to the farmer but can leave the customer lacking for variety. It is all a part of the deal. This provides an education to consumers who quickly learn that unlike a superstore, a CSA cannot provide strawberries and winter squash at the same time.

Seasonality can be a beautiful thing. How wonderful to relish the first spring greens and continue to do so until the hot days of summer come bursting through. What would fall be without crisp apples and pumpkins? CSA customers become aware of growing seasons. It is a great time to educate customers, perhaps giving them a list of which items will be available and their anticipated arrival date.

If the CSA does not include a work share, an open house event for CSA members to come and view the farm gives the customers a sense of involvement. This gives the members a better range of knowledge of what goes on at the farm and the place where their food is actually grown. Public relations efforts such as this go a long way in

continued on page 73

FARMER SPOTLIGHT

John Peterson

John Peterson of Angelic Organics, near Chicago.

John Peterson of Angelic Organics is a man ahead of his time. Born into a family of farmers and after withstanding the challenges of the downward spiral of the farm in the 1980s, John reinvented his farm. He started in a new direction and brought it back to a successful operation.

John's vision for his farm was the CSA. Angelic Organics CSA has over 1,300 members! Located outside of Chicago, the operation draws from the metropolitan area. Services, such as home delivery, make the shares attractive to the masses. Customers rave about the variety and abundance of the CSA share.

John refers to his CSA members as shareholders, giving them a sense of ownership from the beginning. As a part of its mission Angelic Organics strives to do the following:

- Build a sustainable farm system that includes the soil, plants, animals, and humans.
- Provide customers with the highest quality products and best service possible.
- Build community among members.
- Build and maintain optimal soil fertility.
- Provide a safe environment.
- Conduct business in a financially responsible manner.
- Monitor performance against standards.
- Conduct all work in a timely manner.
- Conduct all work efficiently.
- Share knowledge and resources with the larger community.
- Provide employees with opportunities for growth, a balanced life, and adequate financial compensation.
- Provide an orderly succession of management.
- Foster research and development.

- Provide the best possible life for farm animals.
- Create and maintain infrastructure that supports the sustainability of the farm.
- Maintain a commitment to aesthetics and beauty.[1]

A separate entity, the Angelic Organics Learning Center, offers unique opportunities for school groups and other visitors to come and partake of activities especially designed to showcase rural life. For many, this opportunity represents the first time ever visiting a farm. Such activities as milking a goat and feeding chickens make for an experience not soon forgotten.

Its public program offerings include workshops on cheese making, soap making, beekeeping, and organic gardening. There are day and overnight opportunities for families and day camps for kids. A calendar of events is available on the website www. learngrowconnect.org.

Stateline Farm Beginnings is a year-long course offered on the farm for those who wish to learn the business of farming. This is a comprehensive course focusing on launching sustainable farm businesses. There are even interactive workshops called Farm Dreams designed to give potential farmers an inside look at a working farm.

Through its on-farm educational programs, urban partnerships, farmer training programs, and farm and food advocacy work, Angelic Organics touches many lives.

John is the author of *Farmer John's Cookbook: The Real Dirt on Vegetables*. He is also featured in the award-winning film *The Real Dirt on Farmer John*. Visit the Angelic Organics website at www.angelic-organics.com.

Sunflowers speak the universal language of happiness and friendship. It is almost impossible to see a field of sunflowers without a smile. Birds and wildlife appreciate them, too.

continued from page 69

establishing a loyal client base. Your farm becomes their farm, in a sense. If you host an open house, give thought to things like restrooms, hand-washing stations, and farm safety. Have family members or volunteers on hand to direct people. Offer water to drink and a snack if you wish. This kind of day is all about celebration. Cut open a ripe watermelon and let the guests enjoy. These events do not have to be elaborate. Chances are the simpler the better and more memorable for those attending. Check your farm liability policy to make sure you are covered for invited guests.

FARM TO SCHOOL PROGRAMS

Having healthy communities and healthy children is an unwritten goal for the world. In the United States, school kitchens once moved away from homemade food to twentieth-century microwavable, chemical-laden chow. We are seeing a trend reversing these decisions and a resurgence of full-service kitchens, where fresh produce is high on the list of priorities. Salad bars, veggie-based meals, and fruit desserts are replacing the previously offered fat-laden fare. When these programs began, the food directors went running to the commercial food distribution companies. While adding fruits and vegetables to the menu was a step in the right direction, the ball was hit out of the park when the Farm to School program began to connect farmers with schools. According to Farm to School program officials, the biggest problem this program is facing is "finding enough growers to satisfy the demand."

Schools usually do not pay retail prices, but they typically pay market price or slightly above. Selling food to Farm to School programs at retail price can be a good outlet for those who do not have the inclination to sell at farmers' markets. (See Worksheet 10, Appendix 1, for suggestions on how to get started.) Schools are offered incentives to purchase food from within their state of service, which benefits local farmers. (Other institutions, such as prisons, have the same obligation to purchase within the state of service if goods are available.)

Of course, the first thing that comes to mind is the liability issue. Schools have food safety plans in place. A grower must meet the requirements of those specific plans, which are individual to each state and school. If you are interested in filling this niche, start out by searching the national Farm to School website (www.farmtoschool. org). This statement from the Farm to School site sums up the program:

> Farm to School is the practice of sourcing local food for schools or preschools and providing agriculture, health and nutrition education opportunities, such as school gardens, farm field trips and cooking lessons. Farm to school improves the health of children and communities while

continued on page 76

Connie Cunningham

Connie Cunningham with one of her geese at the family farm in Morrison Missouri. Connie has raised more than four hundred geese in a year. Courtesy of Sassafras Valley Farm

When Connie Cunningham returned to the family farm, Sassafras Valley Farm, to care for her aging mother, she knew she needed a project. "We had some land and my brother, sister, and I were talking about what we could do with it to bring it into production. We needed it to make an income. We all thought about things we enjoyed, and we came up with the idea of raising geese. We always had a goose for Christmas dinner. It is a high-end product, relatively difficult to find. It seemed like a natural fit for our land and our family."

Connie began the research, learning about the care of goslings, the proper diet, fencing requirements, guard animals; all the nuts and bolts of raising farm animals. She decided upon two breeds: toulouse and Embden for their French and German heritage. A building was renovated for a nursery, and the first year thirty-two geese were raised as a trial. This was successful, and the second year they produced two hundred grass-fed animals and four hundred in subsequent years. Connie added ducks, last year, and is now considering turkeys.

Connie keeps one group of geese as her permanent flock. "They are favorites and all have names. All have very distinct personalities. They are very social and have their own hierarchy within the flock."

The geese are rotated through a series of paddocks, moving every three days to ensure a fresh grass supply. Corn is fed as a part of their daily ration. Connie works from sunup to sunset caring for the flock and their daily needs. She has researched the native pasture base and has worked to reseed it, feeding the geese the diet the settlers here would have nearly two hundred years ago.

People call or email from all across the country, utilizing the farm's website, www.sassafrassvalleyfarm.com, to place their orders. "Peak season is November and December, but we have some geese available all year long. There is a new interest and emphasis, a real value placed on grass-fed operations because customers are willing to pay more when they know how their food is raised," remarks Connie.

Connie's vision for the farm is a work in progress. Soon she will be opening a B & B, catering to those who want to spend some quality time on the farm. The B & B will be beautifully decorated in a Swedish farmhouse style. Connie collected just the right pieces for years and outfitted the cottage with top of the line linens and bedding. Guests will be supplied with the makings for a country breakfast using local foods, and they may purchase take-home treats from the cottage pantry.

Geese abound on the farm and to watch them in their element is a breathtaking sight. Connie offers a once-in-a-lifetime opportunity to commune with nature in this setting. Birdwatchers will be delighted by the natural habitat of the valley, surrounded by Missouri's rolling hills. Learn more about the farm and the B & B on the website, www.sassafrasvalleyfarm.com.

continued from page 73

supporting local and regional farmers.

Since each Farm to School program is shaped by its unique community and region, the National Farm to School Network does not prescribe or impose a list of practices or products for the Farm to School approach. The National Farm to School Network supports the work of local Farm to School programs all over the country by providing free training and technical assistance, information services, networking, and support for policy, media, and marketing activities. Our network includes national staff, eight regional lead agencies ... and leads in all 50 states. We are here to help you get started and keep programs growing!"[2]

AGRITOURISM AND CULINARY TOURISM

We've touched on localism, celebrating the foods of our regions. Agritourism is a relatively new term that describes the event surrounding sourcing local goods. A trip to the apple orchard, a venture through a corn maze, a trip to a strawberry or garlic festival, the annual cutting of the Christmas tree—all are examples of agritourism, a multimillion dollar industry. It is a noted fact that people want to connect with farms, their heritage, and their region. Jane Eckert of Eckert AgriMarketing states:

> Agritourism is basically where agriculture and tourism intersect, as farms and ranches invite the public onto their property to experience the out of doors, the leisure pace, and the healthy and nutritious produce that is only possible when it is fresh picked at the peak of perfection.
>
> Agritourism, one of the fastest growing segments of the travel industry, includes visits to working farms, ranches, wineries and agricultural industries. Agri-destinations offer a huge variety of entertainment, education, relaxation, outdoor adventures, shopping and dining experiences.[3]

Once again the liability issue comes to mind. If you are inviting customers to come to your farm, you have an obligation to keep them safe. Most people who run agritourism enterprises become good friends with their insurance agents. It is important to cover your ass(ets) to the fullest. Incidences, such as a simple step in a hole or an innocent butt by a goat, can literally cost you the farm. Look for insurance agencies that specialize in farm or special event policies. Shop around and compare policies. Hope for the best, but plan for the worst.

Agritourism opens the creative doors to many new opportunities. What about wreath making, floral design classes, basket weaving from natural materials? The arena is only limited by imagination. People gladly pay to learn and take something home. See Worksheet 14, Appendix 1, for help figuring out how to host an agritourism event.

The aging cave at Vermont Shepherd. Each wheel of cheese goes through a lengthy aging and washing during the ripening process. Much behind the scenes work goes into cheesemaking. The aging process is truly what makes the cheese unique.

BED-AND-BREAKFASTS AND FARM STAYS

Running a bed-and–breakfast (B & B) and offering farm stays open the door to your farm. It takes a special person to do this kind of work and allow visitors to tag along or view farm chores. If you are open to this, farm stays can be a lucrative enterprise.

A farmer in my area offers farm stays that allow guests to cook their own breakfast. The fridge is stocked with local bacon, eggs, fresh bread, jams, and jellies, all available for breakfast. Extra quantities are available for sale in a take-home-the-farm pack. Smart! Customers love it.

If you have an extra building on your property that can be renovated into separate quarters for B & B guests, this type of operation might be for you. Check with your local health department regarding licensing for lodging facilities. Zoning laws vary greatly from county to county and state to state. A B & B operation can require a major upfront investment for a suitable building or lodging space, furnishings, linens, and other considerations.

Customers expect a certain level of elegance, even in a rustic setting. High-quality mattresses, bedding, and linens are important to people who seek out B & Bs as their preferred type of accommodation. Such items as good-quality shampoo, soap, and other amenities are expected. (Consider working with a local soap maker to purchase these items or use this as an opportunity to create a sideline of your own.) Plan to provide extra touches, such as books, games, drawing pencils, and paper, to make the guests' stay as memorable as possible. Market this concept to families for reunions and girls' weekends, and for those who are interested in small farming. If you don't have the space for larger parties, cooperate with other B & Bs in your area to accommodate more guests.

Offer add-ons such as chocolates, flowers, local wines, or even a picnic basket packed and ready for lunch. This is another opportunity to highlight locally produced products while providing a one-of-a-kind experience.

Some B & Bs allow guests to "help" with farm chores or take a farm tour. If this is the plan, work out some simple low-risk tasks that guests can accomplish, such as collecting eggs for their own breakfast. Make sure your liability insurance covers this type of activity. If you have a pond, set out cane poles and let guests try their hand at fishing. The more opportunities you offer to make memories, the more guests are not only likely to come back themselves, but also to recommend your facilities to others.

Provide a list of nearby attractions, restaurants, and shopping in case your guests want to know more about the area. Be only a phone call away if they have a question or need.

Although word of mouth is still the best form of advertising, there are a number of advertising services available to B & B operators. Such sites as www.bedandbreakfast.com offer listings of B & Bs for the potential guest.

CULINARY TOURISM

Another aspect of agritourism is culinary tourism. These tourists pay for a culinary experience, such as a farm-to-table meal or a cooking class. Farm-to-table events have become popular all across the nation. Usually a white tablecloth event is transported to the middle of a farm field. Country charm abounds and local food reigns supreme. This is the opportunity to bring in area wineries, distillers, and brewers in addition to farmers, beef and pork producers, and others involved with food production.

These events bring together farmer, chef, vintner, and those who appreciate food. *Local* takes on a whole new meaning when dinner is served in the field, in the most literal sense. These dinners take a great deal of preparation. However, guests are willing to pay for the experience at a higher than usual dinner rate.

To plan a farm-to-table dinner, first secure your site. Remember this is to be a dinner event, so pick the best spot on the farm, preferably away from animal housing. Consider renting a canopy or a large tent used for events and placing it close to utilities. If you have tables and chairs available, get them ready for the big day. Use vintage or white tablecloths. This is your day to feature all that your farm has to offer, so plan the menu accordingly. Find a local chef who is willing to travel and work with the variety of foods available and who embraces the concept of the farm-to-table dinner. Set the menu. Contact a local winery about the event. Invite it to send someone to pour for the evening. Confirm a date and begin your online campaign with your website and Facebook.

Work with the chef to ensure there are adequate cooking facilities. Most of these events are served buffet-style to cut down on needed staff. If you are planning a sit-down dinner, then provide adequate table servers.

Print an explanation of all locally sourced goods. Invite the farmers to come and dine and make them the heroes of the evening! Have them hand out brochures and business cards to entice direct sales from participants.

Consider a rotating series of dinners at various farms in the area. Use local producers to provide meats, cheeses, vegetables, breads, condiments, drinks, and desserts. Plan for success.

COOKING CLASSES

Cooking classes are always popular. Do you know how to roast a free-range, grass-fed goose, make cheese, or bake bread from locally grown and milled wheat? Want to try your hand at beginning wine making, brewing, or jelly making? There are those who are willing to share their skills with you. Culinary classes for couples have become trendy. Expand on date night and make arrangements with nearby hotels or B & Bs to provide a weekend of food and fun. These events are particularly well received around Valentine's Day. A girls' weekend out is also an enjoyable activity. Often kitchens or halls can be rented for a day or a weekend. Go in and set the stage for your culinary adventure.

HOW TO HOST A COOKING CLASS

If you are thinking of hosting a cooking class, here are some tips and questions to consider.

Will you be teaching the class, or will you be obtaining the services of a chef? If the chef is teaching the class, you will need to communicate your desires to him or her, and make sure the chef has everything needed to carry out the class.

Decide upon a firm number of participants and plan to have a little extra on hand. Just in case someone drops something or another small disaster takes place, be prepared. Do not count on anything being in place in the hall or kitchen you use for your class. Bring in everything. If you do happen to use something from the kitchen, leave a note and replace it immediately. Leave the space in good condition.

Will you provide handouts (recommended)? Will you present a PowerPoint? How will information be delivered?

Will the participants take their creations home or consume them on the spot?

Do you have enough burner space, pots, spoons, plates, serving utensils, napkins, beverages, bread, and condiments?

Plan ahead and make up individual bags, including ingredients, recipes, hot pads, and cooking utensils. This will make the setup go smoothly.

Do a dry run of the class with a willing family member. Write down your talking points, and make sure to emphasize safety.

Allow plenty of time for setup and cleanup.

Bring everything but the kitchen sink with you. Don't plan on the facility you are working in to provide utensils or supplies. As you go through the classes, make sure to keep notes on anything that will help make it easier the next time.

Keep a checklist of supplies to bring and to replace for the next class.

Create a mailing list of class participants so they can communicate with you and each other during the classes and afterward.

Advertise in the local newspaper, cooking section, and at gourmet food stores.

WEDDINGS

The wedding industry continues to grow. You may ask what do weddings have to do with farming. Opportunity. Country weddings, similar to the farm-to-table events, have become quite the social event. Weddings in barns, pastures, creek sides, and other country venues are huge. Combine local foods and wines for the reception, and, of course, if you are a cut flower grower, this is a natural fit. Do you have a country setting that screams "wedding?" Take photos, do the first wedding for a nominal price to build a photo portfolio or host a family wedding, then go from there.

Keep in mind that there are rental fees associated with everything from straw bale seating, Mason jar vases, country crocks, and other accoutrements. Other expenses include contracting with other vendors such as local musicians, photographers, and chefs to develop a comprehensive wedding package (add a slight markup to each service). Feature your foods, flowers, and as much of your own farm-produced goods as possible.

Weddings can range from the basic budget-conscious type to the all-out social event of the year. Do consider there is a great deal involved in pulling off a major event. Start with small weddings and grow from there. Expect some added stress dealing with brides and family members on this very big day. You can simply offer a setting for rent or go all the way to the execution of the full event. And each service has an attached fee. Perhaps you have a future in wedding planning!

CUT FLOWERS

Think about adding cut flowers to your vegetable or field crops to make extra dollars at any market. Learn about post-harvest handling (to make the flowers last longer), take a floral design class, and reap the harvest in

continued on page 82

FARMER SPOTLIGHT

Alan and Betty Nolte

When the economy tanked, Alan and Betty Nolte had to revamp their business. They stopped growing shrubs and trees and started growing vegetables, including heirloom tomatoes.

Alan and Betty Nolte began their career as nursery growers. "Our daughter asked for a small greenhouse for a 4-H project. I started dabbling with it and fell in love with growing. That was forty years ago and I've been growing ever since," states Alan Nolte. The Noltes became widely known for their flowers, shrubs, and trees. Their business sustained them for many years.

Then, things began to change about seven to eight years ago. "When the economy changed so drastically, our business did, too. We provided a lot of shrubs and trees for new housing projects. When those stopped, so did the orders for our products. It happened swift and fast."

Alan decided to plant vegetables. "I always wanted to be a farmer." Now some eight years later, business is on an uphill climb, after a total revamping of the product list and methods of production. "You think it would be easy to go from growing flowers to vegetables, but there is a whole new learning curve," Nolte remarks.

With four greenhouses in production, the major products are hybrid and heirloom tomatoes and English-style cucumbers on a year-round basis. Nolte also gardens outside and in a high tunnel. He has mastered the art of seasonal growing. Along the way, the Noltes found they had more produce than markets, so they employed a broker to assist them in marketing their products. Alan Nolte states, "I wouldn't work with a broker if there was a possibility of moving all of the product myself." However, in their case, a broker was deemed a necessary piece of their overall business plan.

"You have to get to know and to trust your broker," states Alan. "All brokers are not created equally."

Because the Noltes live in a decidedly rural area, the broker assists them in accessing clients in the city, some two hours away from their farm. "Our broker works with chefs and restaurants. These are markets that would have been difficult for us to gain access to," says Alan.

"I think every broker should start out by being a farmer. That way they would know all that is involved. No one knows the hours and behind-the-scenes work that goes into farming, except another farmer," says Alan . "You've got to promote yourself to your broker, tell him your story, so he can sell for you. He is your agent."

In addition to their broker, the Noltes sell through a farmers' market and provide goods for an area CSA. They still sell vegetables from their front porch and roadside location as well. The Noltes grow vegetables beyond the usual green beans and carrots. "We've mastered growing heirloom tomatoes indoors. They said it couldn't be done!" Adding to the vast array are a variety of lettuces, Brussels sprouts, kohlrabies, rutabagas, bok choy, and other lesser-known vegetables. "People like trying something new."

"We started out in 1973. After all those years I am still trying to figure it all out. I look at what worked that shouldn't have, what didn't work that should have, and go from there. I don't know of any other business that keeps you learning and researching after all those years. There is always a new product, a new technique or variety to keep it all interesting."

continued from page 79

a literal sense. Cut flowers enhance everything from a farmers' market booth to a CSA box. Old standbys like zinnias, peonies, sunflowers, sweet peas, and stocks are hard to resist. Develop a standard-size market bouquet to add a splash of color to your operation. A fresh bouquet from the market is hard to resist. Be prepared to be asked about weddings. Develop a portfolio and a price list before the questions arise. Whether cut flowers are your primary product or you grow them as an aside to compliment your farmers' market stand, flowers always draw attention—and customers.

FARM STANDS AND U-PICKS

There are two primary types of on-farm enterprises. The farm stand is a tried-and-true method of selling goods from the farm—on the farm.

FARM STANDS

Farm stands range from stalls under canopies to full-blown retail outlets. There is something about going to the place where the food is grown that can be very appealing to customers. Suddenly, it is an adventure and an event to visit the farm stand. A particular favorite is a stand that has a brisk business in the fall with apples and cider. Many of these types of outlets offer additional goods and services, capitalizing on their semi-captive audiences. Stands may feature a petting zoo, straw bale maze, or other child-friendly activities. Photo ops abound. People enjoy places where they can make memories.

Of course, at a farm stand, the producer can charge full retail price for their goods. No wholesaling, no ordering in advance or other arrangements. Cash on the barrelhead is still a good way to do business.

However, many farm stands are manned with an employee who will accept checks or even credit cards, depending upon the setup.

A popular stand is in Vermont Shepherd at Westminster, Vermont. Here, a small building is designated as the on-farm store. Artisan cheeses, woolens, soaps, maple syrup, and other goods are sold on a self-serve basis. It is encouraging that this type of unattended commerce can still take place. "We have never had a problem," states David Major, the owner of the farm. Major states, "We started this little store as a convenience to our customers. I like being able to offer our cheese on a local basis."

Other examples of farm stores range from apple orchards with fruit prepicked and ready for sale to fully stocked minimarkets. Many of these types of stores bring in other products, such as jams and jellies, to augment their stock. Sales of those types of items are the icing on the cake and often help pay the utilities and possibly the wage of someone to mind the store. Make sure your jams and jellies are from an approved source and labeled to be sold to the public.

Respect for the customer and the customer's time is important no matter how large or how small the retail space. Make sure if you post regular hours that the store is open and stocked. Someone might make a second trip if they find the store closed the first time. It is doubtful the customer would make a third attempt. If you are inviting customers onto your property, check on your liability insurance and find out if additional insurance is recommended. It only takes one problem to create a monster. Check with the local health department to see if a license is required. Generally, a merchant's license is required by the county or city in which business is to be conducted. Sales tax may be required. Will you have an unattended cash box? Will you man the store, or will it be self-serve? These are all questions to be answered before business can begin. Take a look at some farm stands in your area and talk to local shopkeepers to see how their business runs. Assess the pros and cons and then make educated decisions.

U-PICKS

U-picks are another popular on-farm enterprise. In this manner of conducting business, people actually come to the farm and pick their own produce. This might be blueberries, strawberries, apples, peaches, or even asparagus. U-picks are often favorites of those producers who have high-value crops, such as berries, that are labor-intensive to pick. However, it is not simply a matter of opening the farm and allowing customers to come and pick at will. There is some education involved; be it minimal, it is still very important. Let customers know where they are to pick. Are there certain sizes or colors of fruit they should pick and some they should leave behind? Show examples of each. Mark off rows with flags or other markers so customers know where they are allowed to go. Are they to pick a whole area? People tend to cherry-pick, so make sure the subprime fruits go along with the prime! A well-run pick-your–own-asparagus farm sends a measuring stick out with its u-pickers. If the asparagus is not as high as the stick, then the instructions given are to leave it behind for the next picking session.

Advertise your u-pick with flyers posted around town. Signs at the farm will draw potential customers in.

It is important for people to know they are guests on your farm and that there are rules to follow. Imagine several families all showing up at once. Plan for a crowd. Will you supply restrooms, hand-washing stations, and other amenities? If not, make a note on the flyers you post: "No restroom facilities available." People will come prepared if they know what to plan for.

Of course, in a u-pick someone will need to man the operation and give instructions as well as handing out baskets or buckets, then weigh the produce and collect the money. It is helpful to collect email addresses to build a mailing list. One blast with collected names can let your customers know of your next u-pick date or other event.

TRADITIONAL FOOD STORES, BROKERS, AND CHEFS

Sometimes farmers find that they have a knack for retail. The farm stand morphs into a larger and larger enterprise until the farm stand can no longer handle the amount of business. At this point, the farmer must make decisions regarding scale, priorities, bankroll, and other game-changing determinations. A number of small farmers have gone into full-blown retail with mixed results. Often, the co-op model comes into play with several farmers joining together to form a business. Within this model, work and expenses are shared among the farmers/owners. Profits are based on volume and inputs, as they are within the farm stand setting. In other words, the more the farmer provides and sells through the store, the larger his or her share of the profits. This model has worked in a number of operations. It is a way to enter a retail space without the burden of all of the expenses associated with a store.

The store location is a huge part of the quotient. You know the old saying: location, location, location. It certainly rings true in this case. One thing to be considered, whether you are asking customers to visit your farmers' market booth, your farm stand, or a retail outlet, is that you are requesting a change of behavior. Let's face it: It is mighty convenient to stop at one store and get 99 percent of what is needed. When we introduce more stops on the route, we have to make it worthwhile. Out of loyalty, a few customers will seek out their favorite farmers for a while. As life gets busier, though, it is harder to make stops at specialty shops. So if you are considering a retail space, make it worth the customers' while to stop. Of course, bread, milk, and eggs are a jumping-off point. If you as the farmer cannot provide these products, find others willing to join your venture. Of course, licenses, permits, and the health department all need to be involved. It is not a simple process. Worksheet 11, Appendix 1, will help you figure out how to start up a retail business.

Some foods may be processed outside of a commercial kitchen. However, when it comes to selling through a third party, the rules change in most situations. Breads, cookies, jams, jellies, etc. then require preparation in a commercial setting. Eggs must be sold with a license, meat must be processed at a certified facility. There are many hoops to jump through.

Then there are utilities, insurance, rent, and wages if you have employees. Equipment, fixtures, and day-to-day supplies are also to be considered. Quite a list of expenses. Do the math and make sure that you will come out ahead before going this route. Of all the marketing opportunities discussed in this book, this is by far the most involved and labor intensive. Work through the business plan and see where the pencil takes you.

WORKING WITH BROKERS

Brokers can assist the farmer in accessing a broad range of clients. They have made the inroads with chefs, restaurants, food suppliers, and other markets. Their business is to help you sell your products, for a fee. Some brokers add their fee on top of your asking price. More commonly, brokers ask for their going rate. If you have a glut of a particular item, then a broker can certainly be of help. However, chances are if you are up to your eyeballs in watermelon, others are too, so establish a relationship with a broker before the need arises. See Worksheet 12, Appendix 1, for suggestions on beginning and maintaining a relationship with a broker.

Consider working a broker into your overall business plan. With a broker's help, you will enter more markets and have a broader reach for your sales. Many brokers are situated in large metropolitan areas and source foods from neighboring countryside farms. This is a helpful service for those who do not wish to challenge the city or don't know where to go once they get there. Brokers, indeed, can be helpful business partners. You will need to find out how the broker wants goods delivered. Bunched?

> With a broker's help, you will enter more markets.

Packaged? Select? Work to fulfill the specifications. Some brokers purchase by the piece, others by the pound or case lot.

The broker services his or her clientele. Don't forget your broker is the middleman, and if he looks good, you will, too. He is providing delivery service to multiple buyers, which takes valuable time. Remember, you are paying for those services.

WORKING WITH CHEFS

If you have the means to establish and service chefs, this can be another alternative income for your farm. It does mean making a good deal of calls and potentially delivery stops. However, chefs typically appreciate the specialty vegetables only a local farmer can and will provide. These items, such as baby vegetables, the more unusual crops, greens, and other seasonal items will make a chef dance around the kitchen! The hardest part of developing this relationship is getting your foot in the door. Call after lunch, between lunch and dinner. Bring samples, business cards, and any marketing materials you may have about your farm. Be authentic. It's okay to look like a farmer who is cleaned up and going in to town. Before you enter this kind of relationship, make sure you have the time to handle the business. Chefs require prompt deliveries, hate shortfalls, and depend on you to deliver—in more ways than one. Worksheet 13, Appendix 1, is a to-do list for farmers who are working with a chef.

Chefs may request vegetables and fruits of a certain size so they look nice on a plate and make a serving size. Be prepared to cater to this type of opportunity and to put extra effort into giving chefs exactly what they are looking for. If you are willing to do this, a loyalty will develop and you will have a long-term customer. You will have competition, so stay on your toes.

Attempt to work with restaurants in neighborhoods, so you can deliver to specific areas without having to go all over the city. For the most part, chefs will request an invoice, rather than payment on the spot. Most payments are handled through the restaurant's office.

Ask the chef to let customers know they are enjoying the fruits of your labors. Offer table tents or come in and speak on local farmers night. Keep that connection with your customers by keeping your name (and your food) in front of them whether they are coming to your farm or enjoying your food elsewhere.

1 "Farmer John Productions," *Angelic Organics Farm*. Accessed May 29, 2014. www.angelicorganics.com

2 National Farm to School Network. Accessed May 29, 2014. www.farmtoschool.org.

3 Jane Eckert, Teaching the How-Tos of AgriTourism, accessed May 25, 2014, www.eckertagrimarketing.com/eckert-agritourism-what-is-agritourism.php.

CHAPTER 7

OTHER SOURCES OF INCOME

In addition to selling your farm's produce, you can make money by selling crafts and gift baskets or maintaining other people's gardens and farms. To sustain your rural lifestyle, you might have to get creative!

Use your time and expertise to create something that others need or want. For instance, if you can raise worms, you can sell their castings to people who need to enrich their soil.

There has always been a line between the haves and the have-nots. The have-nots often perform tasks that the haves don't care to do. Perhaps it is a lack of time or knowledge, but there is a market for this type of service and people are willing to pay for it.

PLANT AND ANIMAL SERVICES

If you love gardening, consider caring for and planting vegetable or flower gardens for others. This can be a very rewarding experience for all concerned. Post your announcements at local grocery stores targeting high-end neighborhoods. In case you haven't heard, vegetable gardens have become quite fashionable. However, before you break ground make sure the homeowner has checked with the local neighborhood association to make sure gardens are approved for the site. Then proceed with the homeowner by discussing what type of garden they would like, placement, and who will care for the garden once it is started. Some homeowners will ask for 100 percent care for the garden, others will take a role in the watering, picking, etc. Decide ahead of time who will assume each task.

Do you have extra space in your greenhouse? You could custom grow plants for others. Small farmers often purchase plugs, prestarted plants, to put in their field. Contract growing can fill your empty slots and help pay for your own expenses.

Many would like to garden but don't own a tiller or tractor. Set a fee and provide this service. This will not only provide income, but also pay for the equipment.

Learn things others don't know or don't want to know. Learn to shear a sheep, butcher a chicken, trim goat hooves, or other country skills. When someone goes looking for this type of information, help them find you! Post your card or flyer in feed stores, for instance.

FARM-SITTING

Even farmers need a getaway at times. Farm sitter, or caretaker, is another type of service. A farm sitter would typically offer service for a day or a week while the farmer takes a vacation or a needed leave from the farm. In this instance, the farmer would have you come out and learn the ropes. Make sure to take plenty of notes and ask for phone numbers in case of an emergency. If there are animals to look after, get the vet's number and perhaps a close friend or family member of the owner. Write down the schedule, amount, and type of feed, and any other special considerations.

Caretaking is more involved and often requires living on the land. This can be an ideal situation for someone who wants to be in the country but cannot afford to purchase land. Caretakers are assigned certain tasks and then look after the property as if it were their own. This often involves checking fences, maintaining animals, feeding, and land and farm maintenance. This can be a very nice arrangement with the right match of people.

FARM CAMPS

Entertaining children on summer vacation offers the opportunity for a unique on-farm business. Farm camps have become a popular enterprise. Typically a day camp is offered for a week or two providing the opportunity to learn about activities on the farm.

To offer a farm camp, several things should be considered.

1. Safety: Make sure to scout the area where the camp will be held to minimize risks.

Check with your insurance provider to make sure you are covered for this type of activity.

2. Patience: Are you ready for a hundred questions and lots of little hands in the pot? Check your blood pressure, and make sure you have the personality to deal with children.

3. Help: It takes a village to raise a child and at least two additional people per 8-10 children to keep things flowing in a positive direction.

4. Planning: Plan several activities for each day. Children enjoy "helping" in such chores as feeding the chickens, collecting eggs, etc. Plan crafts and story times with a farm theme. Allow them to help prepare their snacks and lunch. Make up some pie dough and let them roll it out for their very own strawberry pie.

These activities are only limited by your imagination and energy level. Natural walks, bird identification, plant identification, fishing, rock skipping, and some of the joys of country living are great activities to share with children. Start with your grandchildren or neighbor kids and run your ideas past them. They will let you know if something is "lame." Make sure activities are age appropriate. Decide early on what ages you will allow to come to the camp. Will you offer a morning session only or plan to go all day? Plan activities accordingly. Always have an extra or two planned, just in case. Giving children some unconstructed time to enjoy the surroundings is also recommended. Plan a library and place the books in low-hanging tree branches. Set up an art center for those who would like to paint or draw. Think like a kid again!

Craft ideas abound on the farm. Egg decorating isn't just for Easter! Teach natural dyeing techniques using beet and carrot juice for dyes. Assist the kids in building a birdhouse kit. Simple weaving can be mastered by a 6 year old. Take a look at stick weaving and allow the children to make sashes for their farm camp badges. Of course, if there are animals present, a petting zoo is always a favorite. (Have handwashing facilities available). A garden walk and harvest will promote healthy eating. Picking a salad will help children to know where food really comes from.

The time is right for farm camps, as adults learn more about the goodness of farm-raised goods, they want their children to learn along with them. This is an opportunity for those who are not faint of heart!

This is an opportunity for those who are not faint of heart!

GIFT BASKETS

The gift basket industry is a $3.3 billion-a-year industry.[1] The convenience, elegance, and serviceability of this type of gift is always a welcome present. If you are in a town rich with tourists, a take-home basket of local flavors and foods is a wonderful remembrance on the way out of town. Local jellies, sauces, chocolates, wines, and other favorites

This Alpine goat on Lazy Lady Farm is part of one of the best working herds in Vermont. Owner Laini Fondilier is known for her goats and her great cheese.

are sure to please. Perhaps a local potter makes a special imprinted mug or plate to add to the basket? The whole idea begins with the basket and then grows to make it seem customized for a particular buyer. If refrigeration is available, add cheeses, meats, or locally grown fruit or vegetables.

Begin the basket by choosing an attractive base. A market basket, flower or herb basket, or small wooden crate is ideal. Line the base with brightly colored fabric or excelsior, and then begin with the tallest item. Place this item either in the very back of the basket or in the center, then build around that item. Layer in cheeses, crackers, mustards, a spreading knife, apples, and a few wrapped chocolates. When the basket looks full, cover it with cellophane or plastic shrink wrap. Fasten a nice bow on top and call it a wrap. Baskets are the perfect way to highlight the products your region is known for. If you go to a farmers' market, set up a booth with small crates full of artfully arranged vegetables.

Visit with the local chamber of commerce. Often this organization sends a welcome basket to a new company expressing interest in relocating. Provide that service. Check with the local florist to see if you can work together to provide their gift baskets. During the holidays, set up a booth at a local mall and custom fill baskets for shoppers. The possibilities are endless.

WEB DESIGN AND WEBSITE MAINTENANCE

Let's face it. Many farmers still reject many electronic devices and yet farms are well represented by a website. If you have the talent to take photos, organize them, write text, etc., then you have the opportunity to build a successful career of designing and maintaining websites. Post your flyer at the local farm stores (along with some business cards), attend local farm shows or events and set up a table, and go to the area farmers' markets and do the same. Decide upon a fee structure, either a set fee or hourly rate. Meet with the farmer and gather information such as the history of the farm, the message they would like to convey, and whether there will be a shopping cart attached. Then teach the farmer to maintain the website or go back on a scheduled maintenance date to update the website. After the initial visit, most of the maintenance work can be done online and offsite. If you have enough of these types of accounts, you can build a clientele to the point that this can be your primary business.

PHOTOGRAPHY

Photography is in great demand. For those gifted with the lens, a farm is a dream come true. Every website requires photos. Flyers are better with photos. Family photos are

continued on page 94

FARMER SPOTLIGHT

Kim Carr

Kim Carr shoots a self-portrait with Mr. Brockman, her bull, in the background. Kim has found her niche in farm photos.

Kim Carr studied hard to earn her BS in Agriculture. She secured her dream job and was moving along life's path until her job was eliminated with barely any warning. Little did she know at the time, but it was a blessing in disguise. "I was working for corporate America and I did not want to go back!" states Kim. So she looked to her twenty-acre farm to sustain her. "I figured I'd rather be poor working for myself than poor working for somebody else. What did I have to lose?"

Kim has always been enthralled by photography. She has a natural eye for capturing a moment. She began taking pictures of her own farm animals and making note cards, offering them for sale along with her eggs at the local farmers' market. "I experienced a little success at first, enough to be encouraging. My family was behind me. And I started to do a few different markets." Kim got her cards into a farm-related retail store and began keeping a regular inventory there. "That inspired me to make more calls. I knocked on doors and expanded the card line along the way." Kim began with twenty-four different photos, including her chickens, goats, mules, and other farm subjects.

"My cards are all based on rural America, life on the farm. I found a gap in the market. There were zebra cards and elephant cards, but I never saw a guinea or a turkey! No one else was doing that. Agriculture is such a part of my life. I want to share that with people."

As time progressed, people began to tell Kim she should consider herself an artist as well as a farmer. She began to investigate galleries and art shows and soon was producing her work in larger formats, particularly matted framed prints and canvas. She is now a frequent exhibitor at many galleries in Missouri.

At last count she had an inventory of over 9,000 cards with three hundred images. "I've always loved note cards and I always wanted to be a farm girl. This combination of art and vocation validates me in that aspect. I hope to send a message, to raise awareness of farming, of old barns and things that people take for granted. Even with a small note card, people take home a frameable piece of art."

Kim credits social media with helping her succeed. "I believe it has helped people find me, and it has helped me highlight the work of other artists. I enjoy promoting other people." Wearing many hats from promoter to artist to packaging manager and shipper, it is more than a full-time job. And then there is taking care of all the animals on the farm! Kim states, "I like social media, you can do it at home in your PJs in the easy chair!

"I sell some of my bigger pieces at shows and that is always nice. The cards are what drive the business. It is good to know that people still take the time to write." Kim has observed that most of her customers are women over forty. "In some ways I am selling nostalgia. Its things people remember from their childhood or their grandmother's house." Kim also sees her cards as a consumable product in that once the card is sent, customers come back to buy more. "Cards are a billion-dollar industry."

As Kim showed her work, she would frequently be asked if she would do a photo shoot of people or pets, and at first she declined. Along the way, she decided to take a stab at it, and now she is frequently asked to take family photos and those of treasured pets. "I know I'll never have a studio, nobody will come out here to Timbuktu! So what I do is go out and capture lifestyle photos on farms. People are comfortable in their own environment."

Kim states she is never one to put all her eggs in one basket. In addition to her wholesale/retail card line, she offers fine art prints in shops and galleries, and she does art shows and exhibits along with the photo shoots. She has also done the photography for two books. "I think it is important to develop different sources of income," she states. She is currently working on a children's book. She plans to turn to Kickstarter to investigate funding.

When Kim is asked about her photography equipment, she giggles. She holds up her iPhone, a Cannon, and a Fuji pocket camera. "This is it! I don't know a thing technical. I just point and shoot." She obviously has an inborn gift. "I don't want to go to a photography class. I just like to take pictures," she states. It obviously works.

When asked about her goals, Kim states she hopes to be financially secure, making a living from the things she loves most; she wants to provide for her animals and experience personal satisfaction. "I so enjoy the people I get to meet, the flexibility of my schedule, and to capture a moment in time with my camera." Kim's work can be found at www.kimcarrphotography.com.

continued from page 91

great in a farm setting. Farmers are always thinking about how to add to the bottom line. A farmer friend of mine has made her living by photographing her own farm animals. She developed a line of note cards that centered upon taking photos of her own animals. She started selling at a local farmers' market and then moved on to art shows and galleries. Kim Carr is now nationally known for her photography work.

WRITING

If you are a farmer, you know things that others would like to know. You didn't get where you are without a lot of bumps in the road, learning what to do and what not to do. There is real value in that. Take a walk through your favorite bookstore and check out the magazine rack. Look at all those farm, home and garden, art, and mechanics magazines, all awaiting content.

HOW-TO ARTICLES

Writing how-to articles is a good place to begin your writing career. The next time you start a seed, make notes about what you do. Write down the sequence of steps to follow. Then explain why you do what you do. Take photos to document each step. Type your notes into your computer and start to write as if you were instructing someone to do exactly what you did. Explain the finer points, the wherefores and whys. Allow the pictures to assist in the instruction and storytelling process. Write your first piece for practice.

After you have written your first article, revise it. Read it out loud and see how it flows. Are all the questions answered? If not, make notes and go back to include pertinent details. Check your grammar, spelling, and accuracy.

Then look up a magazine that you enjoy reading and that features the types of topics you might like to write about. Go to its website and look up the writers' guidelines. Most magazines request a query. A query is more or less a synopsis or proposal of the article you would like to write. Do not send your entire article at this point. Explain who you are and why you are the one to write the article. Briefly describe your expertise in this particular area, any previous work you have done pertaining to the proposed topic, etc. This is the point where you are selling yourself and your proposed article. State whether there will be photos available to go with the story. It is always easier to sell a story with photos than without. If you provide photos, the editor doesn't have to go looking for them. Be passionate and bold in your statements in the query. Speak in certain terms, such as "when I write this piece" rather than "if I write this piece." This is as much for your benefit as the editor's. Send your query in through the contact page on the website.

Sometimes you will get a quick return and a message stating the editorial calendar is full. Don't lament this, it happens frequently. Hopefully that message will be signed by someone at the magazine. If it is, you now have a name and an email address. Write back and say thank you for the consideration. Ask if you could be advised of the next opportunity and what kinds of articles are needed. This will give you a slight advantage. Immediately gather your thoughts for your next piece and query that contact again. Let them know you can deliver the article they need and that you would be happy to send it to them on spec. This way they are not obligated to buy the article, but they can read it and see if you have the right tone, knowledge, and skill to write the type of article they are looking for. Many times the editor will respond and say, "This isn't quite what we are looking for. Could you add a recipe?" The answer would be, "You bet I can!" Be willing to bend, especially on your first few pieces. Once you get one article published, then you can query with new authority and say, "I have written for _____ magazine." Site the issue and date and the title of the article. Keep a running list of your articles in print or available online and build a portfolio.

Most people say they cannot write. True, some do not have the gift of putting pen to paper but most people can tell a story. Write your article as if you were sitting and speaking with someone. Use vocabulary you would normally use. Include bits of your personality in the article. Don't over-think it. Write the piece, and then leave it. Go back to it the next day and review, then leave it again. Read it out loud. Read it to a family member. If you need to start over, do so. Keep your original notes and write the piece again. Every day you write, you will have a different frame of mind and a slightly different voice. If you are writing about farming, then your tone may differ on days when the goats got out and lead you on a merry chase before you sat down to write! Timing is everything. Often the time to reflect is at the end of the day when the chores are done and peace is once again reclaimed.

STORIES

After getting the hang of the how-to format, try writing about a particular incident or a day you would like to remember. Or even one you would like to forget! Writing a nostalgic or humorous anecdote is another approach to take. People enjoy reading about farmers and farms, farm animals, farm kids, and more. There are several magazines that buy this kind of copy.

RECIPES

Next, consider recipes. This type of article can encompass seasonal vegetables, a theme of a particular season or holiday, ethnic recipes handed down from the family archives. What about an article all about eggs? Green eggs, blue eggs, the breed of chicken who lays them, photos of that bird, recipes for all kinds of egg dishes. Take one idea, one topic, and run with it. Then go through the steps listed above. Read it out loud, edit, edit, edit! Make sure it has a good flow and tells the story as you intended.

CHILDREN'S BOOKS

Children's books are another genre to explore. Take a trip to the library and pick up some children's books. Note the colorful pictures, the simple text, and the way the story is told succinctly and clearly. It may look simple at first, but when you sit down and try to write a story with few words and a big impact, it is much harder than it looks. Some people seem to have this voice and it comes naturally. Once again, there are opportunities to write for children's magazines. Take a look at www.writersmarket.com to see what kinds of markets are available.

Don't make the mistake of thinking you have to be a bona fide expert to write. Your firsthand knowledge of how to do a particular thing is a skill you have developed over time. Pass it along and learn to believe you can write what you know.

---

1 "Gift Basket Business Statistics," New Business Ideas and Concepts. July 4th, 2009. www.newbornrodeo.com/2009/07/gift-basket-business-statistics.html.

CHAPTER 8

MARKETING AND YOUR CUSTOMERS

Clean, fresh sweet corn is always a market favorite. To be successful, you need to learn what your customers want.

Remember to grow extra pumpkins and squash for autumn: customers will use them for decorations.

As the producer of high-quality farm-raised goods, the farmer/grower is the immediate expert and, therefore, the best salesperson available. Granted, selling takes some practice and personality. However, if you are passionate about what you do, that will assist you in selling your farm goods. You must believe in yourself and the quality of the products you are bringing to market. Maybe your goals did not include marketing your products. If you want the highest return on your labors, direct marketing is definitely the way to go.

Customers expect professionalism. We've discussed planning your market booth and your attitude and approach to selling. The average consumer will spend less than $30 a week at a farmers' market. That is not a great deal of money to go around. So how can you capture as much of that $30 as possible? Be educated about your industry. Read current magazines to learn about food trends. Eighty percent of your business will come from 20 percent of your customers.

It is proven that people who shop at farmers' markets shop based on relationships. Consumers want to hear your story, to be comfortable with you. Remembering your customers will drive sales. Call them by name or remember what type of vegetables they enjoy. Anything you can do to personalize the experience will come back to you in a positive return. One thing that farmers can offer that a grocery store is lacking in is individualized customer service. Remember most people are three generations away from the farm, and they are looking for ways to reconnect with the land. Have a story ready when people ask a question. Be prepared to be a bit of a performer. People want to know what makes you different. Why are your vegetables (or eggs or meat) better than the booth next door?

Making a positive first impression will bring that customer back to you week after week. Consumers see you as their food supplier and food safety expert. They want to know they can trust you. By and large, people who shop at farmers' markets are more likely to cook and more likely to have family meals. Their value system is different from that of the average consumer.

PUTTING ON THE MARKETING HAT

How can you increase that $30 being spent at the market? Upsell. This is a technique used in marketing, offering the customer additional goods to go along with what he or she has already committed to purchase. For example, someone buys a big juicy homegrown tomato. "Would you like some lettuce to go with that? Perfect for BLTs." Many customers will respond positively and agree to the suggested item. There is a difference between being pushy and practicing suggestive selling. Learn where that line is.

Most farmers have a favorite crop. Is it one that you can sell? The advice is to grow what you can market, rather than market what you can grow. Think about liver. Maybe you can offer the best liver in the business but for some reason liver is never a huge seller. No matter how high the quality or how good the taste, there is simply a limited market for liver. So you probably don't want to set out to be in a business that solely promotes liver. It's the same way with vegetables. Some things move and others simply don't. Find out what you can count on week after week and center your growing around those products. Different colors, textures, aromas, and flavors appeal to all five senses. Keep that in mind when planning your garden.

Customers typically buy two things.

1. Good feelings (healthy lifestyle, healthy kids, weight maintenance, etc.)
2. A solution to a problem (What's for dinner? How to get the kids to eat healthy, etc).

So when you sell to a customer, consider those things. They are getting a good feeling first, by being at a market. They are there because they want to support their local farmers. Help them solve their problems. Suggest kid friendly foods, offer recipes to use vegetables in healthy but tasty dishes. The more you can assist them, the better. People are willing to pay more for things they value. Price is different than value. Price actually has little impact on overall sales. In the long run, perceived value is what is being sold.

When you talk about your farm can you emphasize certain attributes? Are you green? Are you organic? Are you certified humane? These terms go a long way in attracting specific consumers. Take a look at your packaging. Is it minimal, recyclable? Do you have a strong social mission? All these are talking and selling points.

Consumers are driving the change in the current market. Have you noticed even television ads are finding the way to bring farmers into the living room? Jelly ads take you to the fields, soda ads take you to the ginger fields, even car commercials are shot at farmers' markets. Food is such a part of our culture. In fact, food is culture, and we can enrich our lives by experiencing new tastes and foods from other cultures. If you will note, the market is a melting pot of races and origins.

Simple observances can stimulate sales. For example, most people go to the right when they enter a market booth.

continued on page 102

FARMER SPOTLIGHT

Karen Davis and Miranda Duschack

Karen "Mimo" Davis and Miranda Duschack enjoy a moment in between harvests. The two make up
Urban Buds: City Grown Flowers in St. Louis, www.citygrownflowers.com. Courtesy of Urban Buds

Happenstance placed Karen "Mimo" Davis and Miranda Duschack in the right place at the right time. The two learned of a derelict property for sale in the heart of St. Louis. The property consisted of four city lots and a 1950s-era glass Lord and Burnham greenhouse, still intact. The two deliberated for several weeks, working out the potential pros and cons, going back and forth on whether they should, or should not, purchase the property. In the end, they decided to make the purchase. The two became proud owners and began planning their urban flower farm. Urban Buds: City Grown Flowers was born.

Both women brought years of farming experience to the project. Mimo has been a flower farmer for twenty-two years, and Miranda has been growing vegetables and raising bees for eleven years. They combined their skills to lay out a business plan for the farm and soon began turning up the soil to bring their vision into being. "The land we purchased had always been a farm. In fact, it is home to the oldest wood-frame house in St. Louis. We can dig in the dirt there without worrying about anything buried beneath the soil," remarks Mimo.

On their four lots, approximately one acre of land, Mimo and Miranda are growing over ninety different varieties of flowers for their cut bouquets, including sunflowers, zinnias, celosias, echinaceas, larkspurs, delphiniums, lilies, and stocks. "It takes a lot of variety to make interesting bouquets week after week," states Mimo.

The women were cautious in their planning, interviewing local florists and learning about the demand and need for specific varieties even before they purchased the property. They have also put much thought into which flowers they will plant to ensure a continuous supply for their markets. Currently, they grow in the greenhouse, a high tunnel, and outside directly in the soil.

While flowers are still seen as a luxury item, Miranda states, "Even in our present economy, there are still people who see flowers as good medicine, as life improvers." Understanding their customers has been important in developing their overall business plan and pricing schedule. "We know that one segment of our market is younger people who are into self-care. They go to yoga classes and see flowers as an investment in themselves. We like to cultivate these young buyers, and we know as their incomes increase, they will continue to buy flowers, perhaps in larger quantities," states Miranda.

Then there are the older buyers, over forty, who are looking for varieties their grandmothers might have grown. "These customers enjoy lav-enders, sweet peas, flowers with a touch of nostalgia," remarks Mimo. "People come to our booth at the farmers' market and see flowers they have never seen before. The flowers sell themselves. Our most popular items are our arranged market bouquets. We offer these at various price points."

In addition to participating in Tower Grove Farmers' Market, Urban Buds runs a weekly route to area florists. "We load the flowers and bring them directly to the florist. Yesterday I took a small bouquet into a florist to show them what was coming in. Who takes flowers to a florist? It was a little sample of what we will soon have to offer and a gesture of goodwill. Very well received," states Mimo.

At present, the women divide their market into several segments:
- Farmers' market: Least pressure
- Wholesale to florist: Higher pressure
- A frequent-buyers card: Adds pressure
- Events: Most pressure

"We realize we are selling more than flowers. In our business we are selling a fantasy. We provide the product, some information, a bit of entertainment, and even a little theatrics. Customers see we love what we do and that we could not do it without them," states Mimo.

"Our farmers' market stand is our retail store from 8:00 to 1:00 every Saturday morning. It's a pop-up shop. We set it up so people have to walk into it to see the array of flowers. When people walk in, they know they are in our space. We create an ambiance within that space, and we have certain rules we set for ourselves, such as no eating in the stand, no drinks on the table, no sitting. We use tablecloths to create an attractive display along with tiered multidimensional display pieces. It is a full wall of flowers."

The partners acknowledge social media as a part of their overall marketing plan. "It is huge. We use Facebook and Pinterest. We are in the process of building a website. This will allow us to keep in contact with our customers and keep them in touch with the farm.

"We have hired certain people to do skilled labor for us. We have an accountant, an attorney, a website creator, a graphic designer, carpenters, and others who are experts in their field. We are flower growers, and we want to focus our energies in that direction," states Miranda.

"Being a farm owner is a different shift. You have to be a farmer, marketer, designer, PR person, and front man. But we love it all. When you put a smile on people's faces, it makes it all worth it. We are grateful our passion is appreciated."

Sometimes it is very hard to remember the customer is always right.

2. Freshness (Bring only your best to the market)
3. Variety (think of the five senses)
4. Service (be welcoming, make a personal connection, acknowledge children and repeat customers).

Some growers do so well at the farmers' market that they hire staff to cover all the bases. This can mean your staff would do additional markets or assist with a very busy market. If you employ people to represent you, make sure they carry your mission with them. Instruct employees on presentation, demeanor, and salesmanship. Remind them that they are your front men and they should present themselves accordingly. Go over the market setup, procedures, and storytelling with them. Everyone should be working toward the same goal, moving goods. Remind them to spiff up when they go to market. It's okay to wear jeans and a plaid shirt, just make sure they are clean and add a professional air to the booth. Don't wear last week's wardrobe. Clean clothing presents a positive first impression.

About midway through the market season, make it a point to go to a market where you are not a vendor. Take note of attitudes, presentation, merchandising; and give a good mental critique. It will open your eyes to things that may be lacking at your own market, and you will go back with a refreshed attitude about all the things you are doing right.

What to do with a difficult customer? You know there will always be those incidences where you simply cannot make someone happy. Offer to replace the item they are complaining about with something

continued from page 99

Wouldn't this be a good spot for your higher end items? Invite the customer into your booth with the configuration of your tables. Tilt baskets toward customers with product spilling out for a waterfall effect. This notes abundance and choice. Most products should be placed at a height falling between the knees and elbows. Use no more than a four-foot table for easy reach. Consumers are somewhat predictable in their behaviors. Remember, your booth makes your first impression. Go as far as to set up a mock booth at home, and then view it from the outside in. Make notes and changes appropriately. This is serious business. Grocery stores pay big bucks to marketing firms to display wares at top advantage. Take time to analyze your presentation and the overall message of your display.

Taking credit cards can increase sales by 20 to 30 percent. Do you carry cash? Many people do not. Make it easy for them to shop. They will spend more with a card than with cash. In addition to convenience, customers seek:

1. Quality (Bring only your best to the market)

else of higher value. Give them a gift certificate or something to pay them for their trouble. Apologize. Do the best you can to make amends and then let it go. Advise your staff to do the same. Empower anyone working in your booth to make decisions about handling a difficult customer. Then back up your employee if necessary. Sometimes it is very hard to remember the customer is always right!

Consider taking a marketing class. They can be fun, and you will learn a lot about consumer behavior.

GETTING TO KNOW YOUR CUSTOMER

After you have decided upon your marketing venue or venues, identifying your customer is an important step in marketing your products. If you are at a famers' market each week, you will begin to see a trend in particular ages, incomes, and lifestyles. If you are observant, you will pick up on these things. Make it a point to fully take in your customers and note facts about them.

There are certain means of tracking customers. Facebook provides important demographic tools. You can do your own marketing surveys by handing out cards with a few questions on them to see who is shopping at your booth. Better yet, ask the market (as a whole) to conduct customer surveys at their welcome booth. The idea is not to be intrusive but to gain knowledge to assist you in your overall marketing strategy. Knowing who your customers are and discovering information about their shopping habits will assist you in growing and stocking your booth.

The US Postal Service offers a direct mail service. With this type of mailing you can target specific neighborhoods. If you mail out a flyer with a coupon using different colors for different neighborhoods, you will have an idea of where your customers reside. If you want to draw from a larger area, target your marketing to the expanded area with a second direct mailing.

We can make a few assumptions about people who seek out high-quality food, be it through a market, a CSA, or a farm store:

• They are health minded.
• They are community oriented.
• They are willing to go the extra mile to source locally produced goods.
• They want to support local farmers.
• They are concerned about food safety.

CHAPTER 9

ORGANIC CERTIFICATION

Consumers will pay more for organic produce: that is, food produced according to the rules of the USDA's National Organic Program.

If you sell less than $5,000 in products per year and you follow the "final rule," you may call your products organic *without being certified, but you can't use the USDA label.*

Organics have been around for quite some time. In the 1970s, customers were happy to find anything organic, so people of that era were dubbed the granola generation. Traditional health food stores were hip, eclectic stores that attracted colorful customers, new moms, and slightly aging hippies. Some of these stores still exist today, but many have been replaced by the slick, upscale versions, national chains that supply organics to the mainstream marketplace. These national chains have been responsible for introducing the exotic, the hard-to-find, and, yes, the organics to the foodies of the world.

Conventional grocers, recognizing this trend, now offer an organic selection within their aisles. It is quite common to see a good variety of organic fresh produce and packaged goods in the local grocery store. Although this mind-set has been a long time in coming, consumers are placing increased value on food produced without chemicals. Improvements have been made in setting a standard for these goods. No longer can the term *organic* be applied without certification. So what qualifies a product as being organic? According to the USDA's National Organic Program (NOP), "Organic is a labeling term that indicates that the food or other agricultural product has been produced through approved methods. These methods integrate cultural, biological, and mechanical practices that foster cycling of resources, promote ecological balance, and conserve biodiversity. Synthetic fertilizers, sewage sludge, irradiation, and genetic engineering may not be used."[1]

We hear the term *certified organic* used a great deal these days. What exactly does that mean? "*Certification* allows a farm or processing facility to sell, label, and represent their products as organic. Any organic operation in violation of the USDA organic regulations is subject to enforcement actions, which can include financial penalties or suspension/revocation of their organic certificate."[2] If you are a producer and have a desire to become a certified organic operation, what is the process? To become certified, you must apply to a USDA-accredited certifying agent. You will be asked for information, including the following:

- A detailed description of the operation to be certified
- A history of substances applied to the land during the previous three years

- The organic products grown, raised, or processed
- A written organic system plan describing the practices and substances to be used

Here is the organic certification process:
- Producer or handler adopts organic practices and then submits application and fees to the certifying agent.
- Certifying agent reviews the application to verify that the practices comply with USDA organic regulations.
- Inspector conducts an onsite inspection of the applicant's operation.
- Certifying agent reviews the application and the inspector's report to determine if the applicant complies with the USDA organic regulations.
- Certifying agent issues organic certificate.[3]

This certificate is valid for one year. The recertification process involves:
- Producer or handler provides annual update to certifying agent
- Inspector conducts an onsite inspection of the applicant's operation
- Certifying agent reviews the application and the inspector's report to determine if the applicant still complies with the USDA organic regulations
- Certifying agent issues organic certificate

What are the gains of organic production? Higher value, higher prices, and environmental impact. Data available from www.live-the-organic-life.com noted the pricing structure in the table, "Cost of Vegetables". These are retail prices.

One must also look at environmental impacts and the overall promotion of health and lifestyle changes associated with organics. A healthier, forward-thinking population can only be an asset to humankind.

COST OF VEGETABLES

Organic
Carrots:	$1.99/3 lbs.
	.89/lb.
Broccoli:	$2.49/lb.
Celery:	$ .99/lb.

Non-organic
Carrots:	$1.89/3 lbs.
	.89/lb.
Broccoli:	$1.99/lb.
Celery:	$ .89/lb.

Another term to consider is the *final rule*. This is the short name for 7 CFR Part 205 or the organic rule. The final rule officially went into effect on October 21, 2002. See the official document at www.ams.usda.gov/AMSv1.0/NOP. Be prepared; it was at last count 137 pages in length.

Note: A producer selling less than $5,000 in products per year and following all aspects of the final rule may call their products *organic* without being certified, but they may not use the USDA label.,

1 www.ams.usda.gov/AMSv1.0/NOPAccreditationand-Certification

2 www.ams.usda.gov/AMSv1.0/NOPAccreditationand-Certification

3 "National Organic Program," US Department of Agriculture, Agricultural Marketing Service. Last modified December 31, 2012. www.ams.usda.gov.

Seedless
Dills !

CHAPTER 10

FOOD SAFETY

You can save the bounty of the season—and sell it—with canned vegetables and fruits. Check with your state's agriculture department to learn the rules about selling canned foods.

A wholesome safe food supply is one of the reasons people seek out locally produced goods. There is a perceived promise behind buying from a local producer that the food has been produced and processed in the safest means possible. Food safety begins at the farm level. Good Agricultural Practices (GAPs) have been implemented at farms across the country. These practices include the regulation of water quality and safety, manure handling and application, wildlife and pet care, worker sanitation, and post-harvest handling.

A food safety plan begins with a self-assessment of the farm. This means looking at the farm from sowing to harvest, from packing to distribution. Chemicals, water supply, and all aspects of maintaining the crop should be considered. Where are the risks? To learn how to write your own food safety plan to visit www.gaps.cornell.edu. The following points are some places to begin to consider as you venture into fresh produce growing and handling.

SANITATION

One of the primary sources of food-borne illness is the lack of basic sanitation by workers. Establish hand-washing procedures and proper facilities on the farm. If you have workers, train them on proper hand washing. This may seem elementary, but it is a huge step in the right direction. Make sure toilets are properly stocked and clean. If workers are ill, they should not be working with produce.

AVOIDING CONTAMINATION

Chemical additives to soil, insecticides, or herbicides should be used according to the manufacturer's direction. Harvest only after the specified time has passed. If manure is to be applied to the growing area, it should be well aged. If animals are housed in the area where produce is to be grown, make sure the runoff is not directed toward the growing area. Wildlife and pets can also be a source of contamination. Care should be taken to restrict both from the growing area. If an area is contaminated with animal feces, this area should not be harvested.

When harvesting, select for quality. Place produce in plastic bins that have been sanitized. "Field packing, if possible, is the most efficient and safe way of packaging produce. Farmers who select, sort, trim, and package produce in the field will greatly reduce the chance the produce will be damaged."[1]

One of the most important rules of post-harvest handling is that produce should be cooled. Consider on-farm cooling units or make a cooling room by installing a CoolBot. This is a device that works with a room-sized air conditioner to bring the temperature down to a colder level, transforming an insulated room into a cooler. This is an affordable means to build a serviceable cooler. Make sure storage rooms, coolers, and transport vehicles are clean and sanitized on a regular basis.

Some crops should be washed; others can simply be brushed free of soil. Use potable water for all produce washing. If you are packing for a farmers' market, use sanitized plastic bins or clean disposable cardboard boxes to transport the produce. If you are selling to a broker or retail outlet, follow the recommended packaging by the buyer.

Proper post-harvest handling ensures a safe food supply and a salable product. Records should be kept of who is in charge of ensuring the GAPs are maintained. Even on a small farm, it is important to have a go-to person. Volunteers should be trained on the expected practices.

In order to sell fresh goat cheese, you'll need to follow federal and local food safety rules.

BETTER PROCESS CONTROL SCHOOLS

Chapman University	www.chapman.edu/bpcs	714-997-6566
Cornell University	www.foodscience, cornell.edu/cals/foodsci/ extension/extensions-calendar.cfm	315-787-2248
Louisiana State	www.lsuagcenter.com	225-578-5207
Ohio State	www.foodinsutries.osu/edu/services/courses-and-trainings-BPCS-English	614-292-7004
Oregon State	www.oregonstates.edu/dept/foodsci/extservices/ect_index.htm	541-737-6483
Purdue University	www.ag.purdue.edu/foodsci/extension	765-494-7220
Rutgers University	www.cpe.rutgers.edu/food	732-932-9271, Ext 622
Texas A & M	www.aggie-horticulture.tamu.edu/commercial/food_processor	979-846-3285
University of Alaska	www.seagrant.uaf.edu/map	907-274-9691
University of Arkansas	www.uark.edu/ue/foodpro/Workshops	479-575-4450
University of California Davis	ONLINE COURSE	530-752-5901
University of California Davis	www.bit.ly/1Sq1Kpw	530-752-5901
University of Georgia	www.efsonline.uga.edu	706-542-2574
University of Hawaii	www.manoa.hawaii.edu/ctahr/pacific-afsp.?page_id=127	808-956-6564
University of Massachusetts	www.umass.ed/foodsci	413-545-1017
University of Nebraska	www.fpc.unl.edu/workshops	402-472-2819
University of Tennessee	www.foodscience.tennessee.edu	865-974-7717
University of Wisconsin-Madison	www.foodsafety.wisc.edu	608-263-0482
Washington State	www.foodprocessing.wsu.edu/programs.html	509-335-2845

A field map assists in identifying the area the produce was grown. As harvest takes place, mark the bins with the number of the field. This is important; in case there is a problem, only the produce from that area would have to be discarded. If there is no identification system, the whole lot of produce would have to be destroyed. Note the picking date along with the field number for ID purposes. With this information, there is a traceback method if there is ever a problem.

Each situation is unique, so these are but a few highlights of the overall planning process. It is strongly advised that you develop your own food safety plan and implement GAPs on your farm.

CANNING

Value-added products are raw products with added ingredients or special processing that makes them worth more than they were in their original state. Examples include jam, jelly, beef jerky, sausage, goat milk soap, and cheese. If you are looking for ways to capture the full potential of a particular product, then perhaps value-added should be considered. Some states have special rules and regulations for actual food processing, but such homemade goods as jams and jellies can frequently be made in a home kitchen. This rule varies greatly from place to place, so be sure to check with your local health inspector before starting this enterprise. Food products, such as salsas and other canned goods, fall under the category of processed foods or non-acidified foods and must be prepared in a certified kitchen. Often the person doing the processing must have passed a certified food

processing examination.

The Food and Drug Administration's (FDA's) regulations in 21 CFR 108, 113, and 114 became effective May 15, 1979. These regulations are designed to prevent public health problems in low-acid and acidified canned food. Those making food that meets the following definition should attend an acidified food course.

According to FDA's 21 CFR 114, acidified foods mean "low-acid foods to which acid(s) or acid food(s) are added; these foods may include, but are not limited to, beans, cucumbers, cabbage, artichokes, cauliflower, peppers, salsas, some sauces, and fish, singly or in any combination. They have a water activity greater than 0.85 and have a finished equilibrium pH of 4.6 or below." Exemptions include acidic food, repacked acidified foods, fermented foods, carbonated beverages, jams, jellies, preserves, and refrigerated foods.

These FDA regulations also apply to low-acid canned companion animal food. Individuals requiring certification should attend a Better Process Control School. This school is specifically designed for individuals with little or no food science or food safety background. The school includes hands-on basic training in laboratory techniques and skills.

A Better Process Control School equips attendees with a scientific understanding of thermal processes and strategies of pathogen control, first and foremost *clostridium botulinum*. By law, each processor of these types of foods must operate with a trained supervisor on hand at all times.[2]

All commercial acidified and low-acid canned food processors located in the United States and all processors in other countries who process acidified or low-acid canned food products for export to the United States must register with the FDA. Processing plants must also submit process filing forms containing scheduled process information for each acidified and low-acid canned food produced and must meet all other requirements of the Federal Food, Drug, and Cosmetic Act and the Fair Packaging and Labeling Act. Importers, wholesalers, distributors, brokers, etc., are excluded from the requirement to register.

If you would like more information on these Better Process Control Schools, contact one of the institutions on page 112 for additional information.

PROPER PRODUCE STORAGE

If you are a fruit or vegetable grower, this section is especially for you. Naturally you harvest your produce at the peak of ripeness and freshness. You take great care in selecting the best of the crop to take to the market. There are ways to keep your produce at that peak. It all relates to post-harvest handling.

Remember your produce is alive and breathing! When the plant is harvested, its natural supply of nutrients is cut off. So then it becomes your job to contol temperature, humidity, light, and air.

There are five natural elements of a plant's life support system:

1. Respiration (breathing of carbon dioxide from air)
2. Temperature
3. Moisture
4. Light
5. Nutrients

Of these all but nutrients can be maintained after harvest. All produce breathes. If it is packed too tightly in a box, the air supply will be restricted and spoilage will occur. Selection of boxes or produce cases (boxes with holes) will allow air to reach the packed items. Different produce items have different respiration or breathing rates. The faster the product breathes, the shorter the shelf life. The rate of respiration is a factor that can be controlled by chilling the product.

For example: Strawberries, a highly perishable product, breathe ten times as fast at 70 degrees as they do at 33 degrees. Stored at room temp instead of chilled, their shelf life decreases from three days to seven hours.

The higher the respiration rate the more care should be taken to keep the produce chilled. However, too much cold can harm such crops as hard shell squash, avocados, and other fruits. Learn which temperature produces the best results this can take some trial and error. The chart opposite will help in understanding temperature and humidity requirements.

POST-HARVEST HANDLING

Proper storage conditions—temperature and humidity—are needed to lengthen storage life and maintain the quality of harvested fruits and vegetables.

Fresh fruits need low temperature and high relative humidity to reduce the respiration and slow down the metabolic process. See Appendix 6.

Humidity is also a concern in maintaining

RESPIRATION VALUES OF PRODUCE

Extremely High	Very High	High	Moderate	Low	Very Low
Asparagus	Artichokes	Raspberries	Apricots	Apples	Nuts
Broccoli	Green beans	Strawberries	Bananas	Citrus	Dates
Mushrooms	Brussels sprouts	Cauliflower	Cherries	Grapes	Dried fruit
Peas	Avocados	Lima beans	Peaches	Figs	Kiwi
Spinach			Nectarines	Garlic	
Sweet corn			Plums	Onions	
			Cabbage	Mature potatoes	
			Carrots	Sweet potatoes	
			Lettuce		
			Peppers		
			Tomatoes		
			New potatoes		

high-quality produce. This humidity can be added by misting, spraying produce with water, or adding ice to tubs and placing the produce on ice. Proper packaging for highly perishable products will also help maintain shelf life. Some products, such as asparagus or cut flowers, can be displayed standing in water.

In essence, by chilling produce and reducing respiration, we are putting the produce to sleep. By lessening the respiration rate, we are preventing the loss of nutrients, avoiding the loss of weight due to drying out, and maintaining shelf life.

Ethylene is produced by plants as they ripen. This gas speeds up the ripening process. If it is allowed to build up in the air surrounding ripe products, it can cause over-ripening. Remove wrappings from ripe products. If you want a product to ripen, cover the product and allow it to ripen at room temperature. Ethylene can be a help or a detriment.

Plan your harvest around your markets. If you have a Saturday market, begin cutting Thursday evening. Waiting until Friday morning is even better. Then take the produce to your cold room to remove the field heat as quickly as possible. If you do not have a cold room, consider building one. Hydro-cooling is another option. Spray or immerge the produce in cold water. The simplest way of hydro-cooling is dipping the produce into a tank of cold water. The water can be chilled down with ice.

At the very least, head to the nearest shade tree and allow the field heat to evaporate. The instant produce is removed from the stem, the deterioration process begins. All we can do is slow that process.

Your customers may also ask how to store their purchases. Be aware of optimum temperatures so you can provide accurate information. Tell them the goods are at the peak of ripeness when they are purchased (if that is the case) and that they should be

Plan your harvest around your markets. Here, just-harvested onions are drying out.

consumed within a few days for the best taste, texture, etc. Remember, your customers expect you to be the expert.

Present your goods in the best manner possible by trimming off leaves that are decayed or wilted and discard those. Keep your trimming area sanitized and clean. Use a clean sharp knife. Throw away trimmings to avoid the transfer of bacteria. Sort your produce according to size. Discard any with splits or bruises.

Limp produce can be revitalized with crisping. Say you have a market on Friday and have some leftover lettuce you would like to sell on Saturday but the lettuce is looking a bit bedraggled. Take the lettuce into a sanitized area and fill a clean sink with cool (not cold) water. Re-cut the stalk/stem to expose new flesh. Once the lettuce is introduced to the water, it will begin to reabsorb moisture and become vital again. Soak for 3 to 5 minutes, then set it in a cooler to crisp. Allow the product to crisp for 45 minutes to overnight, noting the appearance as vitality is restored.

Pack produce into clean boxes. They don't have to be new but they should be free of debris, void of any rodent dropping, bugs, or mold. Produce should not be packed too tightly or too loosely in containers. Either extreme can cause bruising. Air flow will also be restricted if the produce is packed too tightly.

Packaging can also help you provide

a good product. If allowed, customers will sort through lettuce, picking and choosing their favorite salad greens, leaving the others behind. A bag of mixed greens will sell well, be convenient, and save some hassles. Leave one display bag open, so customers can see what they are getting, then twist tie the rest closed. Sell the greens for so much a bag or so much per pound. For corn on the cob, shuck one ear and lay it on top of the rest. Discourage customers from picking and poking your corn! The more people can see what you have to offer, the less they will damage the goods. Tomatoes and strawberries are delicate. Package strawberries in cartons and tomatoes in open baskets so customers can reach in and get what they want. Bunch onions, radishes, and carrots for best sale. If you are selling to a wholesaler, ask about case packing and get the requirements.

As a farmer, you put a lot of effort into getting your product to harvest. Good post-harvest handling methods make all the difference between a sale and a loss. Don't forget to include your branding efforts on boxes, stickers, and bags, whenever possible.

WILD PRODUCTS AND SAFETY

In addition to value-added or processed foods, there are other means to make a living from the farm. Have you considered agroforestry? In this instance materials are grown or collected from the farm for less traditional uses. Examples of this are wildcrafting and collecting non-endangered species of plants for use in the floral industry. Leaves, vines, pods, seeds, and other natural specimens have great value within the floral industry. In some areas of the country, Spanish moss can be gathered; in other areas, grapevine and other woody ornamentals are plentiful. Check with your state's Department of Conservation to learn which plants should not be harvested. Herbs and other natural products can also be harvested and sold to wildcrafting or pharmaceutical companies. Again, the herbs must be collected under stringent guidelines. Contact American Botanicals (www.americanbotanicals.com) for its instructions and buying list. The main point in looking at a farm as a whole is that there may be other opportunities just beyond your doorstep.

Perhaps this is a good time to actually make a map of your farm. You don't have to be an artist. Contact your local National Resources Conservation Agency (NRCS) and ask for a map of your property. Often topographical maps that note soil composition and other details about a property are available. Overhead views are also frequently available. Draw on a copy of the map and note where specific fields are, give the field a number or name and start identifying the area by that name or number. Note which products are produced in specific portions of the property, such as pasture, garden, row crop, pond, forest, and any other specifics. These location names and numbers will help you when you move into the food safety plan for your farm.

1 Wholesale Success, A Farmer's Guide to Selling, Postharvest Handling and Packing Produce, Jim Slama, familyfarmed.org

2 http://fpc.unl.edu/web/food-processing-center/acbpcs

CHAPTER 11

THE BRASS TACKS
OF DOING BUSINESS

French breakfast radishes are a Saturday morning market favorite.

Talk to your insurer about the types of goods and services you plan to offer. Insurers can help you identify risks that you might have overlooked.

How to do business is a broad topic because different locations have varying rules and regulations. We'll take a look at the more stringent of the rules.

GETTING STARTED

If you are establishing a new business, go to your state's Department of Revenue. Ask whether on-farm businesses or farmers' market venues are required to register their business or business name. In your locale, will you be required to collect sales tax? If so, you must register through your state's Department of Revenue for a sales tax number. Typically, a small bond is required to secure this number, then you must report your sales quarterly and pay the associated sales tax. You will collect this tax from your customers at the time you make a sale. The vendor is simply the middleman between the tax collector and the customer. However, it is your responsibility to collect and pay the tax to the Department of Revenue. Businesses have been lost because of failure to manage these funds properly. If you apply for a tax number, be prepared to do the reporting—or else!

Will you conduct business as a sole proprietor, a partnership, a corporation, or a limited liability company (LLC)? Here are some definitions to help you decide which route to go.

A sole proprietorship is as the name suggests. This type of business is owned by one person.

"A partnership is the relationship existing between two or more persons who join to carry on a trade or business. Each person contributes money, property, labor, or skill, and expects to share in the profits and losses of the business."[1]

"Corporations are business entities that are separate from their owners; corporations have shareholders, and the shares may be privately or closely held, or they may be offered for sale to the public (publicly held). Corporations are formed by submitting Articles of Incorporation to the state in which the corporation is doing business. Corporations are taxed separately from their owners at the corporate tax rate."[2]

An LLC is a relatively new way of structuring a business. It is "a hybrid type of legal structure that provides the limited liability features of a corporation and the tax efficiencies and operational flexibility of a partnership. The 'owners' of an LLC are referred to as 'members.' Depending on the state, the members can consist of a single individual (one owner), two or more individuals, corporations, or other LLCs. Unlike shareholders in a corporation, LLCs are not taxed as a separate business entity. Instead, all profits and losses are 'passed through' the business to each member of the LLC. LLC members report profits and losses on their personal federal tax returns, just like the owners of a partnership would."[3]

Visit with your attorney and tax accountant before deciding how to build your business structure.

INSURANCE

The bane of every businessperson's existence is liability. It is impossible to look into the future to foretell the need for insurance. However, we live in a litigious society. Keeping that in mind, one must plan for the worst (or the maximum amount you can afford). Liability begins on the farm. Do you have employees? Do you sell food to the general public? Do you invite people to come to your farm for special events? Do you have a farmers' market stand? Do you sell goods to chefs or brokers? Do you sell goods to a school? Any and all of these situations (plus many more) put you at risk. Basically, walking out the door and breathing in air can create a situation in which you might be sued these days. Let's say neighbor Tom is assisting you with your vegetable harvest and Tom gets hurt. Something as simple as that can subject you to a lawsuit. We are all aware that we live in a blame-based, guilt-implicating society. Injury equals money and that money must come from somewhere. It will either come out-of-pocket and into predicted ruin or through your liability insurance policy.

Insurance people are not the bad guys in this scenario. In fact, some will assist you in assessing your risks, helping you identify

Provide more than adequate supervision for children.

particular flaws in your operation. Their trained eye can help you to see the things you may simply overlook each day. Some insurance companies supply signage to alert visitors that they are indeed on a farm, and there are potential risks. It sounds silly to have to alert people to potential dangers, but we all know of incidents where common sense was left outside the farm gate. Again, things we see each day and take for granted will not be evident to those visiting the farm for the first or second or even third time. Children pose particular risks. Their quick motions often take place before anyone realizes what has happened. Establish rules for farm visits and go over those rules before anyone is allowed onto the farm. Provide more than adequate supervision.

Talk to your insurance representative about the types of events, products, produce, and other saleable goods you plan to offer. Be honest and open. Insurance is an expense, but this is one of those cases where an ounce of prevention is certainly worth a pound of cure. There are policies to protect your property, policies to protect your goods (in case of a food-borne illness) and other products that may apply to an operation that processes food. Most policies are based on sales figures or volume. Do not think that posting a sign such as Not Responsible for Accidents or having someone sign a waiver will get you off the hook. Think of all there is to lose, and then insure accordingly.

BOOKKEEPING AND BUDGETS

For tax purposes, keep track of income and expenses. Consult your accountant about filing taxes and about items that may be deductible. Your income and expense information is necessary for filing taxes, but it also allows you to see where you are spending or overspending, and where you are making the most profit. Some crops are much more profitable than others. While it is almost impossible to break down your time investment, it is important to note which crops bring in the most income with the smallest amount of effort. Cash crops, such as asparagus, are a tremendous initial investment in time in the beginning, annual weeding and fertilizing keep the crop prolific. However, the going rate for asparagus continues to climb, so if you have the initial time and cash investment, as well as the land to utilize, asparagus could be a winner for you. Garlic is a similar crop that pays off over time. It is beneficial to run an enterprise budget.

An enterprise budget is an estimate of the costs and returns of producing a product, or enterprise. For example, an Iowa corn and soybean producer would be interested in developing both a corn and soybean enterprise budget. Vegetable growers who may have thirty-five to forty different products may wish to develop budgets for their key products, those products that contribute the most to the attainment of producer goals.

Why use enterprise budgets? In economic terms, enterprise budgets help allocate land, labor, and capital, which are limited, to the most appropriate use. The most appropriate use is defined by the person in control of the resources and may be what maximizes profits or minimizes soil loss. The estimated costs and returns illustrated in this publication are based on farm data received from three small farms over a three-year period. Not all crops were grown on each farm each year. The original farm-derived budgets were adjusted slightly to make them more uniform regarding ownership costs, fertilization costs, amount of inputs, and other expenses.

The budgets were developed on a 4 feet by 100 feet bed basis to better represent small farm production. The exception is strawberries, which is on a per acre basis.

Farmstead Cheese from Lavy Dairy, Golden L Creamery of Silex, Missouri. Becky and Hunter Lavy offer samples of their artisan-style raw milk cheeses.

Budget Format

Enterprise budget formats vary. Some are complex. Others are quite simple. Note that the budgets included in this publication are divided into five sections. The first section illustrates the total receipts the enterprise provides on a set unit(s). Records should be kept on both a sales unit (per pound) and land unit (per bed) basis. The second section is the costs of planting and growing the product. These costs are segmented for two reasons. First, these costs are incurred whether a product is sold or not. Once the seed is planted or weeding is completed, it is a sunk cost and needs to be covered from some source.

The second purpose is there is a time delay between pre-harvest expenses and the time the product is sold. These expenses may have to be covered by borrowing or savings or some other source. Therefore, interest on pre-harvest costs should be included as a production expense.

The third section is the harvest component. Note that packaging costs are included with the harvesting activity. Pre-harvest and harvest expenses are combined to equal total variable costs.

The fourth section relates to ownership costs. Each producer owns or controls assets that they use to produce income such as land, machinery, irrigation equipment, and other items. Ownership costs are an allocation to realize some return for the use of those assets. In this example, the land use is $160 per acre. It is assumed that produce is grown on seventy beds per acre so the $160 cost is shared by the seventy beds, or $2.29 per bed. Machinery investment is assumed to be $7.14 per bed or about $500 per acre. Machinery is assumed to have a three-year life so the total machinery investment for replacement purposes is $1,500 per acre. Therefore, a three-acre produce farm would have approximately $4,500 worth of machinery investment that could be replaced every three years.

Freshly baked artisan breads are visually appealing. Take a photo of your goods and post it on Facebook or Pinterest. Experts advise spending at least one hour each day marketing your farm goods on social media.

The irrigation system is assumed to need replacement every three years for a total per acre investment of approximately $240. Total ownership costs are estimated at $10.57 per bed. The last section is the summary of returns. Total costs would be variable and ownership costs combined. The return over variable costs would be total receipts minus total variable costs. The return over all costs would be total receipts minus combined variable and ownership costs.[4]

Running an enterprise budget can be an eye-opening experience. We often hold onto certain crops because we have a connection to them, enjoy growing them, or for other emotional reasons. However, it is important to look at the whole picture to see if there is a true financial gain involved with production. Use head over heart thinking.

Of course, keep track of all your receipts on a daily basis. This can be as simple as having a designated box to drop the receipts in at the end of the day. If you total these up once a day and at the end of the year, you will have an accurate record to file with your taxes. Consider using a program such as Quick Books to assist you with your record keeping. Employ an accountant to assist you with the maintenance of your books. Accountants are the experts in this field. Chances are they don't grow vegetables as their primary income. And chances are you don't do accounting for your primary income. They know the ins and outs of their work, just as you know yours. This is an investment that will pay you back in the long run.

A separate bank account is also a necessity. Track your income against your expenses. It is hard to keep an accurate accounting of either if you are working out of your personal checkbook. Designate this account for farm use only.

FEDERAL ID NUMBER

A nine-digit federal ID number is used to identify a business as a taxpayer. Businesses with no employees and sole proprietorship may use their social security number for tax reporting. Companies with employees must have a federal employer identification number (FEIN). For each business owned by the same person, a different FEIN is required. This number is unique to a business just as a social security number is unique to an individual.

To apply, contact the IRS and ask for the SS-4 form. There is no application fee. Make sure you send it through registered post to guarantee the safe delivery of your form.

For IRS purposes, Schedule F (Form 1040) is the Profit or Loss from Farming form. This document walks you through the steps of listing income and expenses, allowing you to calculate the tax due on your income. It is only as accurate as your bookkeeping, so this is the place where those receipts will become valuable pieces of the accounting pie. A good accountant can take you through the entire process of filing your income taxes.

BRANDING

As previously noted, business cards are one of the best investments a small business owner can make. In addition to having the cards, attempt to brand your farm in every conceivable means possible. Establish a farm name and use it as a brand. Invest in an inked stamp with your farm name, logo, website, or email address. Use it on receipts, bags, tags—any piece of paper that leaves your farm. You never know where those paper tags will wind up, and they are promoting your business.

Consider some of the largest brand names in the country. When you see the golden arches, you know what is coming up. A red-and-white banner on a label immediately makes us think of Coca Cola. These brands are old standbys that were once created as a means to identify a place or a product.

Maybe your farm won't go global, but creating your brand will serve you well. This branding can extend to farmers' market totes and CSA boxes, as well as individual farms. Whatever your organization or business,

Each cheese at Vermont Shepherd is aged just the right number of days to allow perfect ripening.

identify it and make a lasting impression. Be sure to include a phone number and email or web address. Make it easy for customers to reconnect with you. If you want to register your trademark, go back to your attorney and inquire about registration.

WEBSITES AND SOCIAL MEDIA

If you do not have a website, consider creating one or perhaps having a Facebook page. There are also free hosting sites available. An afternoon spent building a website is time well invested and will pay you back in dividends beyond belief. A website (or a Facebook page) allows you to post your hours, available goods, and other communiqués with your customers. Having a Facebook or Twitter feed allows you to communicate sudden changes in your schedule, your availability, or your available goods.

For example, it's an hour until closing time for the farmers' market. You are there and you have an abundance of baked goods that you do not want to take back home. Send out a message on Facebook or Twitter advertising your specials. Facebook photo uploads are a great way to invite people to come and join in on an event. Remind people of where you are and what you have to offer. Send this out to your established mailing lists and ask people to come to the market to support your efforts. "Special today: cinnamon buns, a baker's dozen for $5.00," or something to that affect will go a long way in helping move the product out of your stand and into the hands of appreciative buyers. Tweet when you set up and open the stand. Tweet midmorning about the crowd or the atmosphere. This is a free means of keeping in contact with your buying public.

Pinterest, www.pinterest.com, is also a great way to communicate with those with similar interests. The Food and Drink section is the place to post recipes. Send your clients to these pages for ideas on ways to prepare unusual foods. Instagram, www.instagram.com, is another vehicle to share photos with your customers, friends, and family. All of these social media options can work in concert to assist you in staying in touch with your buying public.

Facebook has tools that can be of help to you. Create a business page, and then take advantage of the features available. You can even schedule posts in advance. Say it is Saturday morning and you are up to your ears in customers or garlic, whichever the case may be. You really don't have time to post something. With a bit of planning, you can use the tools to schedule posts in advance. What a concept! It has been said that it is good practice to post two to three times a day to keep your friends interested in what you are doing. This post-in-advance feature goes a long way toward making that possible. Experts recommend posted at 9:00 a.m., 2:00 p.m., and 8:00 p.m. to coincide with work schedules and end-of-the-day checks. Do remember that Facebook is not a website. Conduct your business on your website. Link your Facebook page to that site. In fact, connect your blog, Facebook page, and website, then send that information to your smartphone.

Facebook can also track demographic information from those visiting your page. This is an invaluable service in learning who your customers are. Track this information to better understand your market. Then consider the age, income, and other pertinent data when planning your stock. Consider this data in your pricing and overall market strategy. Bait your hook to suit the fish!

Consider putting up a YouTube video. Let's say you grow garlic. Shoot a video following the stages of growth throughout the seasons. Most people have no understanding of how garlic is grown. Show the planting of the bulbils, record the progress week by week, add some music to the background, and tell your viewers about your activities. Follow the garlic progression through the seasons, ending with the harvest, drying, braiding, etc. You have now told the story of garlic and more than likely have attracted a few new customers. Refer your existing garlic customers to your video. Bring a laptop or tablet to the farmers' market and run the video.

A laptop or tablet at the market can offer another means of inviting people (virtually) to your farm. Take still shots or shoot video footage of activities around the farm. Shots of goats playing, chicks in the nest, corn in the field, all provide your customers with that feeling of buying in to your dream. Once, a local newspaper did a story on my farm. It appeared the day before my market, which was stellar timing! The next day a lady came to my booth and said, "I just had to come and buy something to take home. I want to be a part of this somehow." People still love farmers. Bringing a laptop or tablet to the market will accomplish more than giving your customers a view of your farm; it will also provide talking points for you and interesting vignettes to point out. It can be fairly taxing to say the same thing 100 times a day! Having a video or photo album can assist you in developing your spiel, which you will use often. Don't become a robot with a canned message, though. Customers appreciate a genuine spirit. You know the kind only farmers carry with them. Be real.

Do not overlook the power of the blog. Consider starting a blog about your farm. A quick Internet search reveals farm blogs from all over the world, from small backyard operations to major factory farms. Blogging is a means to bring people to your farm, virtually or otherwise. Existing customers and potential ones start to feel a kinship with your farm, your family, your livestock, and the garden. Not to mention the development of an understanding of why you do what you do. Blogs are online diaries, so share what you want people to know and understand about your life, while not being too open with personal details. Some blogs are funny and look at the lighter side of farming. Everyone enjoys a snicker at those "it was one of those days" stories. Somehow it makes people feel a little bit better to know there are others in the same boat. When you post something new, readers will follow your blog. Photos are a great addition to a blog and aid in bringing your readers (potential customers) directly to your farm.

Experts advise spending at least an hour a day on social media outlets. That seems like a lot of time to spend in front of the computer when there are cows to milk and vegetables to harvest, but statistically speaking, this is time well spent.

Try some organic, locally grown, sustainably produced cherries—delicious!

According to a report by Social Media Examiner, 86 percent of marketers indicate that social media is important for their business, up from 83 percent in 2012. It's fairly obvious that social media will continue to have a significant impact on the way marketers and business owners continue to communicate with consumers on a daily basis.

Marketers should give attention to which social platforms help them reach their goals with relevant audiences, whether that's generating sales or greater visibility. Pinterest drove 41 percent of social media traffic to e-commerce sites, and the average shopper referred from the network spent between $140 to $180.[5]

If you have a smartphone, you have an invaluable tool. The phone serves as a time-piece, calendar, app store, and your link to the world. Learn to use your particular phone to the fullest. Imagine being in the middle of your garden and seeing a new pest. You can take a picture, send it to a university entomologist, and have it identified on the spot.

Of course, selfies, pictures live at the farmers' market, and shots from the farm all serve to update and involve your clients, giving them a feeling of inclusion and ownership. You can even use your phone to turn up the heat in the greenhouse! If you have

not purchased a smartphone, you might consider the investment. Once you learn the ins and outs of it, you will wonder how you made it without it.

FOOD CIRCLES, FOOD HUBS, AND FOOD APPRECIATION
FOOD CIRCLES

Food circles celebrate community, farmers, and localvores, bringing all of these groups together to serve a common cause, local food production and appreciation. A description from the Kansas City Food Circle notes,

> "The Kansas City Food Circle is building a community food system in which farmers, eaters, chefs, and grocers know and trust each other. Our network enables us to share our knowledge and experience while we work together promoting the benefits of locally-grown organic and free-range foods. Our primary purpose remains to support farmers in our region who strive to produce delicious, nourishing food, use environmentally-sound, organic methods that restore and preserve the health of the land and who treat their animals far more humanely than industrial animal factories."[6]

The University of Missouri notes:

> "A Food Circle is a new way of conceiving of and organizing our agricultural and food systems. It links the many people involved in food production together in interdependent, holistic ways. When we conceive of our food system as a circle, we acknowledge that we are connected with every other person in that circle through the act of food production.
>
> "Practically, a Food Circle is concerned with promoting the consumption of safe, regionally grown food that will encourage sustainable agriculture and help to maintain farmers, who will sustain rural areas. While the concept sounds simple, it means that we must radically change the way we participate in the act of growing and consuming food."

Somewhere along the way we lost the concept as stated above, "conceiving of our food system as a circle." Undoubtedly, since the industrial revolution and the advent of canned foods, there has been something removed from our food, namely, the farmer. Food circles bring the farmer back into the center of food production. They celebrate regionalism and localism. Often, food circles are built to educate and appreciate consumers and to introduce the farmer. This might take shape as a monthly potluck dinner featuring everyone's favorite dishes, farm-to-table dinners, or community picnics. There is a touch of nostalgia here. What better way to preserve traditional foods of a specific region. If specific dishes and recipes are not held in high regard, they will be lost in our convenience-based world. Food circles can serve as a repository for old recipes and traditional favorites. Why not a food circle cookbook?

FOOD HUBS

According to the USDA's Regional Food Hub Resource Guide, a food hub is a business or organization that actively manages the aggregation, distribution, and marketing of course-identified food products primarily from local and regional producers to strengthen their ability to satisfy wholesale, retail, and institutional demand.

This is another example of people and associations coming together to meet the public demand for high-quality food.

Food hubs may facilitate aggregation and delivery services for multiple farms. They can encompass direct deliveries to schools and other institutions. The hub works to combine the efforts of many, sparing the farmer some of the chores of packing, fulfillment, and shipping. Food hubs might provide insurance, educational

opportunities, and other benefits to those who are members.

The central idea behind the food hub is to help producers approach larger markets. Small producers may not be able to supply the demand, but when food is collected and combined, large orders can be accepted and filled. Hubs offer services that small farmers may have trouble providing, such as refrigeration, delivery trucks, and the infrastructure required to serve larger markets.

"A good example of this is Red Tomato, a nonprofit marketing and distribution organization based in Canton, MA. Founded in 1996, Red Tomato arranges the aggregation, transportation, and sale of a wide variety of produce supplied by 35–40 farmers to grocery stores and distributors in the Northeast. It never physically handles the product sold under its name but instead relies on farmers and contract trucking firms to provide aggregation and transportation services."[8]

SLOW FOOD

Perhaps the best known food appreciation organization is the Slow Food movement founded by Carlo Petrini and a group of activists in the 1980s with the initial aim to defend regional traditions, good food, gastronomic pleasure and a slow pace of life. In over two decades of history, the movement has evolved to embrace a comprehensive approach to food that recognizes the strong connections between plate, planet, people, politics and culture. Today Slow Food represents a global movement involving thousands of projects and millions of people in over 160 countries.[9]

Slow Food envisions a world in which all people can access and enjoy food that is good for them, good for those who grow it, and good for the planet. It opposes the standardization of taste and culture, and the unrestrained power of food industry multinationals and industrial agriculture. Its approach is based on a concept of food quality that is defined by these three interconnected principles:

- **Good:** A fresh and flavorsome seasonal diet that satisfies the senses and is part of the local culture
- **Clean:** Food production and consumption that does not harm the environment, animal welfare, or human health
- **Fair:** Accessible prices for consumers and fair conditions and pay for producers

Slow Food believes food is tied to many other aspects of life, including culture, politics, agriculture, and the environment. Through our food choices, we can collectively influence how food is cultivated, produced, and distributed, and as a result we can bring about great change.

When was the last time you savored a meal? Generally, as Americans we are interested in getting a meal on the table as quickly as possible. However, in other cultures a meal is a celebratory effort. Wines and cheeses, sausages, olives, breads—all come from the neighboring countryside. Perhaps mimicking some of these practices, these moments in time, can once again begin to be infused into our own cultures. We see that happening bit by bit. It all begins with simple, humble foods rather than something out of a box or a bag that is eaten in the car.

Slow Food fosters farmers.

"Conviviality is central to our mission. We are a global community, connecting people to the land and to each other through local projects, educational events, and shared meals. We become catalysts for change by sharing the joy of Slow Food and prioritizing wholesome living over convenience. We champion local, culturally significant heritage foods, customs and recipes—and bring these experiences into farms, markets, restaurants and homes. We teach the next generation how to grow, prepare and share food responsibly."[10]

To find a convivium in your area check the Slow Food website, www.slowfood.com.

LOCAL HARVEST

Local Harvest is a public online directory of small farms, farmers' markets, and other local food sources with an emphasis on sustainability. Participants may sign up for an online store that allows them to sell their goods online, offering a nationwide storefront. Farmers go online and list their products, which are then entered into a searchable database of products and farms. Blogs, events, and websites can be linked through

These Norton grapes are at the peak of ripeness. They're reserved for late-harvest wine.

the site. You can find everything from honey to farm crafts, wool, meats, fresh vegetables, and more.

TROUBLESHOOTING AND ORGANIZATIONS TO KNOW

Every business has its potential pitfalls. Here is some advice to help you prevent or solve problems you might face on your farm:

Your produce doesn't sell: Part of this problem can be eliminated with pre-marketing research. If there is too much of one kind of produce at a market, then perhaps it is time to investigate another market location or another type of market.

The crop is lost: Investigate crop insurance. Check with your area Farm Service Agency for insurance options.

Talk with your peers: Talk to fellow farmers to see what has worked for them, and, perhaps more importantly, what has not worked for them. Farmers are generally good about sharing information with those getting started.

Treat your farm as a business.

Maintain contact with your team.

Avail yourself to educational opportunities. Go to a growers' conference and learn from seasoned growers or university experts.

AGENCIES

There are a number of agencies in place to help small farmers.

SUSTAINABLE AGRICULTURE RESEARCH AND EDUCATION (SARE)

Since 1988, the SARE grants and education program has advanced agricultural innovation that promotes profitability; stewardship of the land, air, and water; and quality of life for farmers, ranchers, and their communities. SARE is funded by the USDA National Institute of Food and Agriculture (NIFA). It is one of the few organizations that awards grants directly to farmers. Search the SARE website (www.sare.org) to view grants that have been successful in the past. SARE also offers a searchable database of projects that have been funded in the past. There is a great deal of knowledge available on this website.

Do you have a particular project you have wanted to experiment with? Recently a cheese maker noted that when she poured whey on her squash plants that she did not see populations of squash bugs. She submitted a proposal to SARE and won a grant to see if her premise was correct. She did the research (for which she was paid by the grant) and proved her theory. The introduction of whey did, indeed, lessen the number of squash bugs on her zucchini plants. The cheese maker provided valuable information that may be useful to others, satisfied her curiosity, and had a successful zucchini crop; all in all, a win-win situation.

Here is some information on the SARE Farmer Rancher Grant Program:

Farmer Rancher Grant Program offers individual ($7,500 maximum), partner ($15,000 maximum), and group ($22,500 maximum) grants for ideas initiated by farmers and ranchers.

Projects may last up to twenty-two months. About fifty projects are funded each year.

A sample of the calls for proposals is available throughout the year, but you should always consult the current call for proposals when applying.

The annual portfolio of producer grant proposals is reviewed and awarded on an annual timeline.

Grants support producers who are protecting natural resources, enhancing communities, and boosting profitability.

Farmers and ranchers are exploring innovative marketing of sustainable agriculture projects in addition to other project topics.

Outreach and networking multiplies producer project results.[11]

The call for proposals for SARE farmer rancher grants typically occurs annually in late August.

THE FARM SERVICE AGENCY

The Farm Service Agency (FSA) is an agency of the USDA. You may be able to get a loan or loan guarantee through FSA's farm loan programs if you are a farmer or rancher who is unable to obtain credit elsewhere to start, purchase, sustain, or expand your family farm. Unlike loans from a commercial lender, FSA loans are temporary in nature, and the goal is to help the farmer or rancher graduate to commercial credit. Once you are able to obtain credit from a commercial lender, the mission of providing temporary supervised credit is complete.

The FSA targets a portion of its loan funds to minorities and women farmers and ranchers through its Socially Disadvantaged Applicants (SDA) funding source. Applicants need to provide their race, ethnic origin, and gender when applying for an SDA loan. If you're looking for an FSA guaranteed loan,

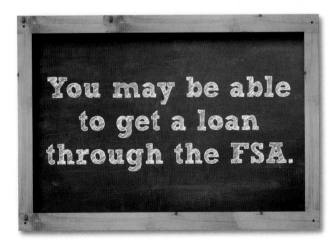

direct operating loan, or direct farm ownership loan, you may qualify through SDA. Contact your local office or USDA service center to learn more about its programs. See a listing of state offices in the Appendix.

THE SMALL BUSINESS ASSOCIATION

The Small Business Association (SBA) offers a number of loan programs for business owners who may not be able to qualify for a traditional bank loan. The application process is done at a participating bank or other lending institution that offers SBA programs. The application is structured so the loan is eligible for an SBA guarantee, which promises to pay a portion of the SBA loan if the borrower defaults.[12]

SBA also offers a microloan program for projects under $50,000.

NATIONAL RESOURCES CONSERVATION SERVICE

The NRCS conservationists provide technical expertise and conservation planning for farmers, ranchers, and forest landowners wanting to make conservation improvements to their land. NRCS also helps America's farmers, ranchers, and forest landowners conserve the nation's soil, water, air and other natural resources. All programs are voluntary and offer science-based solutions that benefit both the landowner and the environment.[13]

UNIVERSITY COOPERATIVE EXTENSION SERVICES

The county agent is still alive and well. University Extension is a division of land grant universities that is dedicated to bringing information from a university directly to individual counties within a particular state, providing research-based information. You may know University Extension best by one of its signature programs, 4-H. However, Extension works in communities each day, assisting farmers; offering community building assistance; nutrition and health information; home gardening advice; and other programs. See the listing of land grant universities in Appendix 3.

THE UNITED STATES DEPARTMENT OF AGRICULTURE

Each state has a Department of Agriculture, which oversees agricultural production and other issues. It is the   federal executive department responsible for developing and executing federal government policy on farming, agriculture, forestry, and food. See Appendix 2 for a list of agriculture departments by state.

Goat cheeses from Baetje Farms in Bloomsdale, Missouri. The Baetjes have put Missouri on the map with their artisan cheese production.

1 "Partnerships," *IRS*, Last updated May 22, 2014, www.irs.gov/Businesses/Small-Businesses-&-Self-Employed/Partnerships.

2 Jean Murray, "US Business Law / Taxes," *About.com*, Accessed May 29, 2014, biztaxlaw.about.com/od/glossaryc/g/corporations.htm.

3 "Choose Your Business Structure," *The US Small Business Administration*, Accessed May 29, 2014,www.sba.gov/content/limited-liability-company-llc.

4 Craig Chase, "Iowa Fruit & Vegetable Production Budgets," Iowa State University Extension and Outreach: Ag Decision Maker, Accessed May 28, 2014, www.extension.iastate.edu/agdm/crops/html/a1-17.html.

5 Brian Honigman, "Social Media Year in Review: 13 Must-Know Statistics from 2013," Entrepreneur, December 23, 2013, www.entrepreneur.com/article/230488.

6 "Our History and Legacy," *Kansas City Food Circle*, Accessed May 29, 2014, www.kcfoodcircle.org/about/about-history/.

7 "Mission: What is a Food Circle?," *Food Circles Networking Project*, Accessed May 29, 2014, www.foodcircles.missouri.edu/mission.htm.

8 James Barham et al, "Regional Food Hub Resource Guide," U.S. Dept. of Agriculture, Agricultural Marketing Service, April 2012, p. 5, http://dx.doi.org/10.9752/MS046.04-2012.

9 Slow Food, "Slow Food: The History of an Idea," Accessed May 28, 2014, www.slowfood.com/international/ 7/history.

10 Slow Food USA, "About Us," Accessed May 28, 2014, www.slowfoodusa.com.org/about-us.

11 www.sare.org

12 "Loans," *U.S. Small Business Association*, Accessed May 29, 2014, www.sba.gov/content/sba-loans.

13 www.nrcs.gov

APPENDICES

The barn at Vermont Shepherd, more than 200 years old, is at the heart of the farm.

APPENDIX 1
Worksheets

Worksheet 1:
SWOT ANALYSIS

STRENGTHS

_____
_____
_____
_____

WEAKNESSES

_____
_____
_____
_____

OPPORTUNITIES

_____
_____
_____
_____

THREATS

_____
_____
_____
_____

SUMMARY

_____
_____
_____
_____
_____
_____

Worksheet 2:
MARKETING PLAN

PROJECT BACKGROUND

_____
_____
_____
_____

PROPOSED OPERATING STRUCTURE

_____
_____
_____
_____

AVAILABILITY OF SUPPLY

_____
_____
_____
_____

LABELING

_____
_____
_____
_____

DESCRIPTION OF PRODUCTS TO BE MARKETED

_____
_____
_____
_____

Worksheet 2 *(cont.)*

EXPECTED PRICING

_____
_____
_____
_____

COMPETITION

_____
_____
_____
_____

PROCESSING

_____
_____
_____
_____

FOOD SAFETY

_____
_____
_____
_____

MARKETING TRENDS

_____
_____
_____
_____

Worksheet 2 *(cont.)*

DISTRIBUTION

_____
_____
_____
_____

MARKETING TACTICS

_____
_____
_____
_____

EXECUTION OF MARKETING STRATEGY

_____
_____
_____
_____

NOTES

_____
_____
_____
_____
_____
_____
_____
_____
_____
_____

Worksheet 3:
TEAM IDENTIFICATION

1. Farmer/producer

_____

_____

2. Advisors (family, friends, others who can offer assistance)

_____

_____

_____

_____

3. Attorney

_____

4. Accountant

_____

5. Insurance provider

_____

6. Other

_____

_____

_____

_____

_____

_____

_____

_____

_____

Worksheet 4:
ASSET IDENTIFICATION

1. Land or garden space. Can be owned, borrowed, rented, or leased

_____

_____

_____

2. Available operating funds, credit, or financing available

_____

_____

_____

3. Knowledge—what skills do you have that will be of benefit to your farming operation?

_____

_____

_____

4. Who do you know that can help? Family, friends, educational opportunities?

_____

_____

_____

5. What are your passions? Growing, working with people, animals, etc

_____

_____

_____

Worksheet 5:
INVESTIGATING A FARMERS' MARKET

1. Land or garden space. Can be owned, borrowed, rented, or leased

_____

_____

2. Does the market appear to be organized and well run?

_____

_____

3. Do the hours of operation fit your schedule?

_____

4. What is the atmosphere? Is there music? Are people smiling? Are they carrying bags of purchased goods?

_____

5. Is there a market master?

_____

6. Does the market operate with a board of directors?

_____

7. Does the market carry liability insurance

_____

8. Are vendors required to carry product liability insurance?

_____

9. Is there a producers'-only market?

_____

10. What is the pricing structure?

_____

Use this information to ascertain if you are in the right place to conduct your business.

Worksheet 6:
MARKET DAY CHECKLIST

_____ Canopy

_____ Tables

_____ Tablecloth

_____ Sign with farm name

_____ Business cards

_____ Display pieces (racks, baskets, bins, etc.)

_____ Price tags or signs

_____ Properly packaged goods

_____ Bags

_____ Calculator

_____ Scales

_____ Money box

_____ Change

_____ Receipt book

_____ Clipboard with pages to collect names for mailing list

_____ Samples and recipes, if desired

_____ Water

_____ Merchant's license, if required

_____ Sales tax chart

Worksheet 7:
HOW TO OPEN A FARMERS' MARKET

☐ Contact friends and neighbors to ascertain interest in a new market.

☐ Schedule an initial meeting to discuss the possibility of a new market.

☐ Contact local health inspector to advise of new market.

☐ Find a location.

☐ Elect or appoint a governing body and market master.

☐ Prepare vendor agreement and establish fees.

☐ Set schedule.

☐ Contact area newspapers and other news organizations for publicity.

☐ Start a market Facebook page.

☐ Obtain liability insurance.

☐ Seek volunteers.

☐ Lay out market spaces ahead of first day.

☐ Establish an information booth at the entrance to the market.

☐ Invite vendors.

☐ Plan for opening day. Make arrangements for water, electricity, portable toilets, and other amenities.

☐ Plan events throughout season.

☐ Have regular meetings throughout the market season as well as a closing meeting for the season.

Worksheet 8:
VENDOR AGREEMENT FORMS

Here are two samples of vendor agreement forms. Customize them for your own use. Add or subtract items that do not apply. Make sure to include issues that specifically pertain to your market.

Sample 1

Market Location: _____

I. General Market Policies

A._____Farmers' Market is managed by the market master, who has authority over all market operations.

B. _____ Farmers' Market is open for operation on _____ beginning_____ .

C. _____ Farmers' Market will open at_____and close at _____p.m. each market day. Vendors are required to stay until closing.

D. The market master has the discretion to limit the number of vendors for each category at registration. The market master reserves the right to allow or disallow all items and vendors of any category at his or her discretion.

E. _____ Farmers' Market is smoke-free.

F. No firearms or weapons are allowed on the market premises.

G. No solicitors are allowed at_____ Farmers' Market.

H._____Farmers' Market does not discriminate against anyone because of race, color, creed, national origin, sex, disability, or sexual orientation.

I. *Vendor* is defined as the actual producer who grows, crafts, bakes, or makes final product to sell. This may be an individual, immediate family member, another producer member of the market, or an employee of the producing individual's farm or operation. No selling of non-produced items allowed. Exceptions must be approved by the market master.

J. All vendors at the _____ Farmers' Market must be local. *Local* is defined as produced or handmade within a_____mile radius of_____ .

K. Due to health regulations, vendors are not allowed to have pets at the market.

L. One space will be made available free of charge for educational activities relating to sustainable agriculture or nonprofit health-related community groups. This space will be made available on a first-come, first-serve basis but must be approved and booked with the market master in advance.

Worksheet 8 *(cont.)*

M. Anyone not complying with the rules of the _____ Farmers' Market may be asked to leave the market.

N. Any exception to these policies will be at the discretion of the market master.

O. Vendors are responsible for compliance with applicable city, county, state, and federal regulations such as (but not limited to):
1. Vendors are not required to obtain city business license.
2. All scales used by vendors must be certified by the Department of Agriculture and the approval seal of Weights and Measures be displayed on all scales. Scales are not required unless sold by weight. Items may be sold by the piece or bundle. Non-certified scales are not allowed, and you will not be allowed to use them at the market.
3. Sales Tax: State, city, and county sales tax must be collected. Each vendor is responsible for his or her own tax collection and reporting.

II. Allowable Products
A. Food Items
1. Fruits and Vegetables
   a) Whole uncut fruits and vegetables only
2. Honey, maple syrups
3. Baked goods (see Part B below for prohibited baked goods)
4. Jellies, jams, preserves
   a) Jellies, jams, and preserves may not be made with artificial sweeteners
5. Wine
   a) Must have appropriate licensure to sell
6. Value-added products*
   a) Examples: soup mixes, baking mixes
7. Homemade candy
   a) Fudge
   b) Brittle

Non-value-added products will not be permitted for sale at the market. This includes the repackaging of spices and herbs. No independent distributors.

B. Prohibited Baked Goods
1. All cream pies
2. All whipped cream–topped pies
3. All whipped topping–filled pies
4. All meringue-topped pies
5. Cream-filled pies
6. Boston cream pies (pudding-filled cake)
7. Cheese cakes or cheese Danish (all cream cheese products prohibited)
8. Pumpkin pies
9. Custard or custard pies
10. Pudding or pudding pies
11. Mince-filled cookies or pies

Worksheet 8 *(cont.)*

12. Flan
13. Goods made from boxed mixes

C. Other Prohibited Products
1. Home-canned items. Example: tomatoes, salsa, BBQsauce, any vegetables, etc.
2. Meat jerky of any kind, unless approved by the_____ Health Dept.
3. All plants

D. Eggs
1. Eggs may be sold if the vendor has proper licensure. This includes a retail license and a dealer's license from the _____ Department of Agriculture.
2. All egg cartons must be labeled with the date eggs were packed, name and address of dealer, or approved dealer identification number.
3. All crates used for transport and sale of eggs must be provided by the vendor.
4. Eggs must be candled and graded prior to sale.
5. Eggs must be transported and kept at proper refrigeration temperature (45°F) through point of sale at the market.
To obtain a complete copy of _____ Egg Laws & Regulations, call the_____ Department of Agriculture, Division of Weights and Measures at_____ or visit its website at_____ .

E. Nonfood Items
Vendors selling crafts do not need overhead protection. The following are allowed for sale:
1. Approved crafts (handmade by vendor)
2. Small animals and livestock (dogs and cats are prohibited)
3. No bedding plants, herbs, trees, or shrubs allowed for sale.

F. Meat Products
1. Meat products are permitted as allowed by the_____ County Health Department. Please call the Environmental Section at _____ .

III. Special Policies for Producers and Food Item Vendors
A. Vendors selling food items must provide overhead protection (when the market is outdoors) and a table for their products. All produce or other food-related products must be displayed either on a table or in a container if at ground level.

B. Producers selling products they refer to as organic must display a sign giving their organic grower's certification and their certifying body, unless exempt from certification due to scale of operation.

C. Producers should clearly separate and label organic and nonorganic products in the same display.

D. Each vendor producing home baked goods must have a sign at his or her table that indicates the products have not been inspected by the _____ Health Department.

Worksheet 8 *(cont.)*

E. Shade structures shall be secured to the ground via clearly marked blocks, sandbags, or other heavy objects in order to prevent damage to products and injury to others.

F. Food safety, sanitation, health permits, and labeling requirements pertaining to the items for sale are required.

G. Honey labeling must contain the statement: "Do not feed to infants under 1 year of age."

H. Prepackaged items must be labeled with the following information (this includes all food products):
1. A list of all ingredients in descending order
2. A statement that indicates the product was not inspected by the _____ County Health Department.
3. Name, address, and phone number of the vendor or person who manufactured the product.

Sample Label **NOTE: If you need help with labels, tell the market master, and we will be glad to assist you in any way.

Ingredients: Flour, eggs, milk, pecans, salt
This product was or was not inspected by_____ County Health Department
Name and address of manufacturer:
Justin Case
555 Apple Lane
Anywhere, USA 11111
Your Phone #

IV. Application and Reservation Policies
A. All vendors must complete and submit to the market master the items below before being permitted to sell at the market each week. The market master must receive all required paperwork before the vendor is permitted to sell at the market.
1. Complete _____ Farmers' Market Application Form. This form is available at

_____ .
2. Proof of vehicle liability insurance.
Each vendor is required to have and maintain liability insurance to operate a vehicle within the _____ Farmers' Market. A copy of the insurance ID card is required with the vendor application.

B. Vendors must be approved by the market master.

C. Vendor signature on the vendor application verifies the vendor has carefully read, understands, and agrees to _____ Farmers' Market rules and regulations. Booth assignments, determination of suitability of items offered for sale, and collection of booth rental fees are the responsibility of the market master.

Worksheet 8 *(cont.)*

D. Each vendor booth will be inspected by the market master on the vendor's first week of attendance, and the market reserves the right to inspect at any time to verify compliance with the health department and market regulations.

E. The weekly vendor fee is $_____ .

F. All fees must be paid by check, cash, or money order made out to "_____."

G. Any vendor writing a bad check will be charged $25 per check and no future checks will be accepted. Bad check fees must be paid in full prior to renting future space.

H. Payments are to be given to the market master.

V. Setup and Pricing Policies

A. Vendors must be set up by_____ . Late arrivals may be turned away. No sales are permitted before the opening time or after ending time. A bell will ring to start and end the market.

B. Vendors must clean up the area around their vehicles and booth before leaving the site each market day.

C. The market master will determine space and parking designations. All vendor spaces will be 10 x 10 feet. Vendors must stay within their designated space. Vendors going outside their space will be required to pay an additional $10.00 per 10 x 10-foot space.

D. Vendors should park in a designated area if unable to park adjacent to their booth.

E. Radical price cutting of top-quality products is not permitted.

F. Signs identifying the name and location of the vendor's business must be posted before sales begin.

G. Signs, boards, tags, or labels listing prices of all products for sale must be posted prior to the beginning of sales.

H. It is the responsibility of the vendor to provide any tables, overhead protection, chairs, change, cash registers, scales, signage, tents, etc., necessary to do business. Some equipment is available for rent on a first-come, first-serve basis.

I. A vendor is required to keep all vehicles, contents, products, and byproducts in the boundaries of his or her assigned stall at all times, no matter how many empty stalls are located throughout the market.

J. A vendor is to keep all spare stock, packing materials, cardboard boxes, and bags in an orderly fashion at all times.

Worksheet 8 *(cont.)*

K. Vendors should provide their own change for customers.

L. Vendors are required to have product liability insurance.

VI. Inclement Weather

A. If the weather is bad, we may have a rain delay until the storm passes, but we will not cancel the market immediately. If the storm is severe and does not pass and is not predicted to pass, we will cancel at that time. If a vendor decides not to come because of the weather, he or she must notify the market master but will not be charged for that week.

B. If the market is canceled prior to_____ , vendors will be refunded their vendor fee for the week. If the market is canceled after_____a.m., no vendor fee refunds will be given.

Sample 2

This market stall lease ("Agreement") is made as of_____ , 2014 (today's date) ("Effective Date") between Owner:
and
Vendor: Your Name: _____
Business Name: _____
Address with zip: _____
Phone: _____Cell Phone: _____
Email:_____

1. **Lease.**
Owner hereby leases to Vendor the stall, as identified below, during the ("Event") for the permitted use on each Saturday during the term, and the Vendor agrees to pay a fee to Owner as set forth herein and perform all other agreements set forth herein.

2. **Market Area.**
The market area is described as located within _____
_____.

3. **Term.**
____ Full Season ($315) May 3 to October 4, 2014, twenty-three Saturdays
____ Half Season ($175) twelve Saturdays of choice:_____
____ Partial ($110) 6 Saturdays of choice: _____
Important—Indicate the first date you expect to attend the market: _____
If requesting two booths, indicate here ___ and include additional payment with check.
Second booth = Discount to 75% of single booth rate.
a. Briefly describe the items you intend to market (Note: allowed items as per guidelines):
_____
_____

4. Fees.

Vendor agrees to pay fees of $_____ for a season permit lasting for a full term, $_____ for twelve (12) selling days, or $_____ for six (6) selling days at the market. Fees are to be paid with contract application, in full, per Vendor's respective choice of term. All checks shall be payable to _____.

5. Refunds.

Owner will not refund any fee payments made unless Owner is unable to provide the stall or space of approximate equal area to the stall. Vendors who fail to set up in a prepaid stall by 7:30 a.m. on any Saturday during the term may forfeit that stall along with any fee paid in advance. Owner then has the right to reassign the stall to another vendor.

6. Event Sale Hours and Cleanup.

Vendor may not arrive or set up any earlier than _____a.m. Vendor must operate at the stall fully set up until ____ p.m. on each day of the event during the term. Vendor shall be responsible for the final cleanup of the stall, including broom sweeping and complete removal of any and all items or other materials used for the sale or display of any and all merchandise no later than ____ p.m. on event day. Sale hours are ____ a.m. to ____ p.m.

7. Safety, Rules, and Regulations.

Vendor shall take all reasonable precautions for the safety of its employees and all customers and visitors shopping in the farmers' market area and around its stall. Vendor shall comply with all applicable laws, ordinances, rules, regulations, and lawful orders of any public authority bearing on the safety or protection of persons or property located on or near the farmers' market area and in or around the stall.

8. Farmers' Market Guidelines.

Vendor shall comply and cause its employees and assistants to comply with any reasonable rules and regulations established by the Owner, including but not limited to the farmers' market guidelines and the _____ rules and regulations (collectively, the "Rules"). Vendor hereby acknowledges receipt of a copy of the guidelines. Owner may change or alter the Rules as it deems necessary or appropriate for the overall good of the farmers' market area.

9. Permitted Use.

Vendor may use the stall only for the sale of those items designated on the attached Vendor & Market Guidelines—Exhibit A, unless otherwise permitted in writing by Owner.

10. Insurance Coverage.

All Vendors must carry personal liability insurance and product liability insurance. Proof of such insurance shall be made available upon request.

11. Indemnification.

Vendor shall defend, indemnify, and hold Owner, _____, any parents, affiliates, principals, agents, and employees of either and both, as well as market management and any other owners harmless from any, against any and all claims

Worksheet 8 *(cont.)*

whatsoever arriving in any way out of the Vendor's acts or omissions in, on, and about the farmers' market area and the activities therein.

12. **Assignment.**
Owner may reassign space in the market area as it deems necessary and in its sole discretion, provided that the Vendor is given a space of equal area for which the Vendor paid in advance. Vendor may not assign this lease or allow others to use the stall without prior written consent from the Owner.

13. **Termination.**
This Agreement shall automatically terminate if Vendor does not timely pay fees or fails to fully comply with any of the terms of this Agreement. In addition, Owner shall have the right to terminate this Agreement upon three (3) days' written notice to Vendor if Owner receives more than three (3) complaints about a Vendor's product, performance, or conduct during the term. Vendor deviation from market guidelines is cause for termination of this Agreement.

14. **Vendor's Exclusive Remedy.**
Vendor acknowledges and agrees that its sole and exclusive remedy under this Agreement against the Owner for any reason shall be to require the Owner to refund fee charges not earned by the Owner. Vendor waives any and all other rights or remedies that might be available in equity or in law, including the right to seek damages whether special, incidental, consequential, or otherwise.

15. **Attorney's Fees.**
If any party hereto shall bring any suit or other action against another for relief, declaratory or otherwise, arising out of this Agreement, the losing party shall pay the prevailing party's reasonable costs and expenses, including reasonable attorneys' fees and court costs. The parties have caused this Agreement to be executed by their duly authorized representatives on the date first above set forth. By signing below, Vendor acknowledges that he or she has read and agrees to abide by the market guidelines.

_____   _____

Vendor Signature   Date

Worksheet 9:
CSA AGREEMENT

The spirit of a CSA subscription is for the subscriber to buy a share of our farm's products for a particular season. This means that you as the shareholder will share with the farmer in both the bounty and the risks associated with farming. By signing this agreement, the member agrees to the stated conditions of risk.

Name:_____

Address:_____

City:_____State:_____

Phone Number:_____

Email:_____

Number of shares purchased:_____

Pickup day and location:_____

Length of season (beginning and end dates):_____

Basic share (vegetables and fruits):_____Add bakery_____

Add meat_____ Add dairy_____

Amount due for season $:_____

Signature:_____Date:_____

Worksheet 10:
HOW TO BEGIN A FARM-TO-SCHOOL PROGRAM

1. Meet with school officials to ascertain interest and promote Farm to School program.

2. Once a relationship is established, it is important to make sales and delivery convenient for the school cafeteria.

3. Obtain liability insurance and follow the school's food safety plan. Become well acquainted with that document.

4. Work with the school cafeteria in menu planning and anticipating crop schedule.

5. Deliver requested goods in a timely manner. Work in a cooperative setting to have one or two growers deliver, rather than requiring the school to deal with a high number of growers.

6. Bill the school on an agreed-upon basis and distribute funds to growers.

7. Maintain close contact with the school. Offer to come and talk to students about farming.

8. Foster this relationship for ongoing success.

Worksheet 11:
HOW TO SELL TO RETAIL STORES

1. Call and make an appointment with the appropriate store department head. Do not wait until you have an overabundance of product. Plan ahead.

2. Take samples of your product, if available.

3. Offer a copy of your on-farm food safety plan and your liability insurance documents.

4. Set a schedule of anticipated delivery with store department head.

5. Ask questions about payment schedule and come to an agreement on payment dates and price for goods sold.

6. Deliver top-quality goods. Ask the store to identify you as the grower in its signage. Promote local production.

Worksheet 12:
HOW TO WORK WITH BROKERS

1. Contact area broker and introduce yourself and your farm.

2. Set up a meeting to discuss potential working relationship.

3. Offer samples, if available. Sell yourself and your products.

4. Set up calendar of anticipated delivery.

5. Offer a copy of your on-farm food safety plan and liability insurance documents.

6. Make deliveries as scheduled.

7. Ask questions about payment schedule and come to an agreement on payment dates and price for goods sold.

8. Deliver top-quality goods.

9. Label your product with your farm name, if allowed. This will carry your name into the public.

Worksheet 13:
HOW TO WORK WITH CHEFS

1. Contact chefs on a Monday to set up an appointment.

2. Take samples of your produce, if available.

3. Discuss your commitment to local food production, quality, and flavor with the chef.

4. Find out exactly what the chef is looking for—something unique, a specific size vegetable, heirlooms? Then commit to grow to the chef's standards.

5. Set up a weekly delivery schedule and make that delivery no matter what. If a chef is to rely on you as a primary source of goods, deliveries must happen as planned.

6. Offer your on-farm food safety plan and liability insurance documents.

7. Ask questions about payment schedule and come to an agreement on payment dates and price for goods sold.

8. Deliver top-quality goods at the appointed time.

9. Ask to be featured on the menu as the grower of the produce. Include table tents or business cards, if appropriate. This will promote your farm and bring customers to you.

10. Grow new varieties of produce for your chefs to sample and experience.

11. Occasionally dine at the restaurant and let the chef know you are there.

12. Send customers to the restaurant.

Worksheet 14:
HOW TO HOST AGRITOURISM

1. Decide upon type of event. (If it is a seasonal event, such as an apple harvest, plan according to the date of harvest. If it is a farm-to-table event, try to avoid the hottest days of the year.)

2. Decide upon date and time.

3. Start to advertise on Facebook, through the local papers, radio, etc.

4. Check with your insurance provider to make sure you have adequate insurance for this type of event.

5. Ask friends and family to volunteer to help or hire a caterer with a staff for dinner events.

6. Arrange for tent, table settings, linens, stemware, etc. Arrange for restrooms and hand-washing facilities.

7. Arrange to work with a local winery or brewery. Many will send staff for an event, for a fee.

8. Work with a local musician to provide music for the event.

9. Plan the menu according to seasonal availability.

10. Send invitations to specific groups.

11. Two days ahead of time, erect the tent and set up tables and chairs. Do a final check-in with the caterer to make sure numbers are in and everything is ready to go.

12. On the day of the event, lay out table linens, flatware, and place settings. Have greeters in place at the appropriate time.

13. Assist catering staff with setup, if needed.

14. Begin cocktail hour.

15. Enjoy the event, you are the star! Make a brief announcement, thank people for coming, then talk about your farm and the significance of the event. Make sure to mention other producers who have provided food, wine, etc., for the evening.

APPENDIX 2
State Agriculture
Departments

Alabama: 5 S. Broad Street, Samson (334) 898-2151, www.agi.alabama.gov

Alaska: 334 4th Ave., Seward (907) 224-3374, www.dnr.alaska.gov/ag

Arizona: 1688 W. Adams Street, Phoenix (602) 542-4373, www.azda.gov

Arkansas: 1 Natural Resources Drive, Little Rock, (501) 683-4851, www.arkansas.gov

California: 1220 N. Street, 4th Floor, Sacramento, (916) 654-0466, www.cdfa.ca.gov

Colorado: 700 Kipling Street, Lakewood (303) 239-4100, www.colorado.gov/ag

Connecticut: 165 Capitol Ave., Hartford (860) 713-2500, www.ct.gov/doag

Delaware: 2320 S. DuPont Hwy., Dover (302) 698-4500, www.dda.delaware.gov

Florida: Florida State Capitol, 400 S. Monroe Street, Tallahassee (800) 435-7352, www.freshfromflorida.com

Georgia: 19 Martin Luther King, Jr. Drive, SW, Atlanta (404) 656-3600, www.agr.georgia.gov

Hawaii: 1428 S. King Street, Honolulu (808) 973-9560, www.hdoa.hawaii.gov

Idaho: 2270 Old Penitentiary Road, Boise (208) 332-8500, www.agri.state.id.us

Illinois: P.O. Box 19281, State Fairgrounds, Springfield (217) 782-2172, www.agri.state.il.us

Indiana: 1 N. Capitol Ave., Ste 600, Indianapolis (317) 232-8770, www.in.gov/isda

Iowa: Wallace State Office Building, 502 E. 9th Street, Des Moines (515) 281-5321, www.agriculture.state.ia.us

Kansas: 109 SW 9th Avenue, Topeka (785) 296-3556, www.agriculture.ks.gov

Kentucky: 100 Fair Oaks Lane, Frankfort (502) 564-4983, www.kyagr.com

Louisiana: 5825 Florida Blvd., Baton Rouge (225) 925-3770, www.ldaf.state.la.us/portal/

Maine: 22 State House Station, 18 Elkins Lane, Augusta (207) 287-3200, http://www.maine.gov/dacf/

Maryland: 50 Harry S Truman Pkwy, Annapolis (301) 261-8107, www./mda.maryland.gov/Pages/homepage.aspx

Massachusetts: 251 Causeway Street, Ste 500, Boston (617) 626-1700, www.mass.gov/eea/agencies/agr/

Michigan: P.O. Box 30017, Lansing (800) 292-3939, www.michigan.gov/mdard

Minnesota: 625 Robert Street N., Saint Paul (651) 201-6067, www.mda.state.mn.us/

Mississippi: 121 N. Jefferson Street, Jackson (601) 359-1126, www.mdac.state.ms.us/

Missouri: 1616 Missouri Blvd., Jefferson City (573) 751-4211, www:mda.mo.gov/

Montana: 211 McKinley Street, Lewistown (406) 538-3489, www.agr.mt.gov/

Nebraska: 301 Centennial Mall S. #4, Lincoln (402) 471-2341, www.nda.nebraska.gov/

Nevada: 405 S. 21st Street, Sparks (775) 353-3601, www.agri.nv.gov/

New Hampshire: 25 Capitol Street, Room 220, Concord (603) 271-3685,
www.agriculture.nh.gov/

New Jersey: 369 S. Warren Street, Trenton (609) 292-5517, www.state.nj.us/agriculture/

New Mexico: 2604 Aztec Road NE, Albuquerque (505) 841-9425,
www.nmdaweb.nmsu.edu/

New York: 10B Airline Drive, Albany, (800) 554-4501, www.agriculture.ny.gov/

North Carolina: 2 W. Edenton Street, Raleigh (919) 733-7125, www.ncagr.gov/

North Dakota: 600 E. Boulevard Avenue, Dept 602, Bismarck, 01-328-2231
www.nd.gov/ndda

Ohio: 8995 E Main St, Reynoldsburg, (614) 728-6201, www.agri.ohio.gov/

Oklahoma: 2800 N. Lincoln Blvd., Ste. 100, Oklahoma City (405) 521-4039,
www.oda.state.ok.us/

Oregon: 1205 5th Street, Tillamook (503) 842-2607, www.oregon.gov/ODA/pages

Pennsylvania: 1307 7th St, Altoona (814) 946-7315, www.agriculture.state.pa.us

Rhode Island: 235 Promenade Street, Providence (401) 222-6800, www.dem.ri.gov

South Carolina: 1835 Assembly St. Ste. 1008, Columbia (803) 765-5333,
www.agriculture.sc.gov

South Dakota: 523 E. Capitol Ave., Pierre (605) 773.5425, www.sdda.sd.gov

Tennessee: 939 Upper Ferry Rd, Carthage (615) 735-0300, www.tn.gov/agriculture

Texas: 1700 N. Congress, 11th Floor, Austin (512) 463-7476, www.texasagriculture.gov

Utah: 350 N. Redwood Rd, Salt Lake City (801) 538-7100, www.ag.utah.gov/

Vermont: 109 Professional Dr., #2, Morrisville (802) 888-4965,
www.agriculture.vermont.gov/

Virginia: 02 Governor Street, Richmond (804) 692-0601, www.vdacs.virginia.gov

Washington: 1104 N. Western Ave, Wenatchee (509) 664-2280, www.agr.wa.gov

West Virginia: 1336 State St., Gassaway (304) 364-5103, www.wvagriculture.org

Wisconsin: 2811 Agriculture Drive, P.O. Box 8911, Madison (608) 224-5012,
www.datcp.wi.gov

Wyoming: 2219 Carey Avenue, Cheyenne (307) 777-7321, www.wyagric.state.wy.us

APPENDIX 3
Land Grant Universities

Alabama: Alabama A&M University, Auburn University; and Tuskegee University

Alaska: University of Alaska–Fairbanks

Arizona: University of Arizona

Arkansas: University of Arkansas; University of Arkansas at Pine Bluff

California: D-Q University, University of California

Colorado: Colorado State University

Connecticut: University of Connecticut

Delaware: University of Delaware; Delaware State University

Florida: University of Florida

Georgia: University of Georgia

Hawaii: University of Hawaii

Idaho: University of Idaho

Illinois: University of Illinois at Urbana-Champaign

Indiana: Purdue University

Iowa: Iowa State University

Kansas: Kansas State University

Kentucky: University of Kentucky; Kentucky State University

Louisiana: Louisiana State University; Southern University, and A&M College

Maine: University of Maine

Maryland: University of Maryland–College Park;, University of Maryland–Eastern Shore

Massachusetts: University of Massachusetts–Amherst;
                        Massachusetts Institute of Technology

Michigan: Michigan State University

Minnesota: University of Minnesota

Mississippi: Mississippi State University; Alcorn State University

Missouri: University of Missouri; Lincoln University

Montana: Montana State University (Bozeman)

Nebraska: University of Nebraska–Lincoln

Nevada: University of Nevada–Reno

New Hampshire: University of New Hampshire

New Jersey: Rutgers; New Jersey Agricultural Experiment Station, Office of Continuing Professional Education, School of Environmental and Biological Sciences

New Mexico: New Mexico State University

New York: Cornell University

North Carolina: North Carolina State University; North Carolina A&T State University

North Dakota: North Dakota State University

Ohio: Ohio State University; Central State University

Oklahoma: Oklahoma State University; Langston University

Oregon: Oregon State University

Pennsylvania: Pennsylvania State University

Rhode Island: University of Rhode Island

South Carolina: Clemson University; South Carolina State University

South Dakota: South Dakota State University

Tennessee: University of Tennessee; Tennessee State University

Texas: Texas A&M University; Prairie View A&M University

Utah: Utah State University

Vermont: University of Vermont

Virginia: Virginia Polytechnic Institute and State University (Virginia Tech); Virginia State University

Washington: Washington State University

West Virginia: West Virginia University; West Virginia State University

Wisconsin: University of Wisconsin–Madison

Wyoming: University of Wyoming

APPENDIX 4
State Farmers' Market Associations

Alabama – www. fma.alabama.gov

Alaska – www.alaskafarmersmarkets.org/

Arizona – www.info@arizonafarmersmarkets.com

Arkansas – www.arkansasfarmersmarketassociation.com

California – www.cafarmersmkts.com

Colorado – www.coloradofarmers.org

Connecticut – www.ctfarmfresh.org

Delaware – www.defarmersmarkets.net

Florida – www.fruitstands.com/states/florida.htm

Georgia – www.n-georgia.com/farmers_markets.htm

Hawaii – www.farmersmarkethawaii.com

Idaho – www.farmersmarketonline.com/fm/Idaho.htm

Illinois – www.ilfarmersmarkets.org

Indiana – www.farmersmarketonline.com/fm/Indiana.htm

Iowa – www.iafarmersmarkets.org

Kansas – www.fromthelandofkansas.com/discover-resources/farmers-market

Kentucky – www.kentuckyfarmersmarket.org

Louisiana – www.louisianafarmers.org

Maine – www.mainefarmersmarket.org

Maryland – www.marylandfma.org

Massachusetts – www.massfarmersmarkets.org

Michigan – www.mifma.org

Minnesota – www.mfma.org

Mississippi – www.mdac.state.ms.us/departments/ms_farmers_market/index.html

Missouri – www.mofarmersmarkets.com

Montana – www.montanafarmersmarket.com

Nebraska – www.nefarmersmarkets.org

Nevada – www.nevadagrown.com

New Hampshire – www.nhfma.org

New Jersey – www.njfarmmarkets.org

New Mexico – www.farmersmarketsnm.org

New York – www.nyfarmersmarket.com

North Carolina – www.ncfarmersmarkets.org

North Dakota – www.ndfarmersmarkets.com

Ohio – www.ohioproud.org/markets.php

Oklahoma – www.okfarmandfood.org/oklahomafarmersmarkets

Oregon – www.oregonfarmersmarkets.org

Pennsylvania – www.pafarm.com

Rhode Island – www.farmfresh.org/markets

South Carolina – www.scfarmersmarkets.org

South Dakota – www.farmersmarketonline.com/fm/SouthDakota.htm

Tennessee – www.tnthe.com/THE/TFMA.html

Texas – www.texascertifiedfarmersmarkets.com

Utah – www.utahsown.utah.gov/farmersmarkets/index.php

Vermont – www.vtfma.org

Virginia – www.vfdma.org

Washington – www.wafarmersmarkets.com

West Virginia – www.wvfarmers.org

Wisconsin – www.wifarmersmarkets.org

Wyoming – www.wyomingfarmersmarkets.org

APPENDIX 5
Bibliography and
Suggested Reading

Aubrey, Sarah Beth. *Starting and Running Your Own Small Farm Business*. Storey Publishing, 2007.

Corum, Vance, Marcie Rosenzweig, and Eric Gibson. *The New Farmers' Market: Farm-Fresh Ideas for Producers Managers & Communities*. Auburn, CA: New World Publishing, 2001.

Ivanko, John, and Lisa Kivirist. *Rural Renaissance*. Canada: New Society Publishers, 2009.

Hansen, Ann Larkin. *Organic Farming Manual*. North Adams, MA: Storey Publishing, 2010.

Mancuso, Anthony. *How to Form a Nonprofit Corporation*. Nolo Press, 2013.

Minnesota Institute for Sustainable Agriculture. *Building a Sustainable Business*. MN: Minnesota Institute for Sustainable Agriculture, 2003.

Perry, Jill, and Scott Franzblau. *Local Harvest: A Multifarm CSA Handbook.*, SARE, 2010.

Rangarajan, Anusuya, Elizabeth A. Bihn, Marvin P. Pritts, and Robert N. Gravani. *Food Safety Begins on the Farm: A Grower Self-Assessment of Food Safety Risks*. Cornell University, 2003.

Wholesale Success. Edited by Jim Slama. www.familyfarmed.org

APPENDIX 6
Food Storage Chart

The table below indicates optimal temperature and moisture conditions for some common fruits and vegetables. Chart reprinted with permission, Penn State Extension and Northeast Center for Risk Management Education.

Product	Optimal Storage Temperature (oF)	Freezing Point (oF)	Optimal Humidity %	Top Ice Accepted	Water Sprinkle	Ethylene Production Accepted	Sensitive to Ethylene	Approximate Storage Life	Comments
Apples	30-40	29.3	90-95	No	No	High	Yes	1-12 months	Chill sensitive stored at 35-40 F
Apricots	31-32	30.1	90-95	No	No	High	Yes	1-3 weeks	
Artichokes	32-35		90-95	Yes	Yes	No	No		
Artichokes, Jerusalem	31-32	28.0	90-95	No	No	No	No	4-5 months	
Asparagus	32-35	30.9	95-100	No	Yes	No	Yes	2-3 weeks	
Avocados, ripe	38-45		85-95	No	No	High	Yes		
Avocados, unripe	45-50		85-95	No	No	Low	Yes, Very		Keep away from ethylene producing fruits
Bananas, green	62-70		85-95	No	No	Low	Yes		
Bananas, ripe	56-60		85-95	No	No	Medium	No		
Basil	52-59		90-95	No	Yes	No	Yes		
Beans, dry	40-50		40-50					6-10 months	
Beans, green or snap	40-45	30.7	95					7-10 days	
Beans, sprouts	32		95-100					7-9 days	
Beans. Lima	37-41	31.0	95					5-7 days	
Beets	32-35		90-95	Yes	Yes	No	Yes		
Beets, bunched	32	31.3	98-100					10-14 days	
Beets, topped	32	30.3	98-100					4-6 months	
Blackberries	32-33	30.5	90-95	No	No	Very Low	No	2-3 days	
Blueberries	32-35		90-95	No	No	Very Low	No		
Bok Choy	32-35		90-95	No	Yes	No	Yes		
Broccoli	32	30.9	95-100	Yes	Yes	No	Yes	10-14 days	
Brussels Sprouts	32	30.5	90-95	Yes	Yes	No	Yes	3-5 weeks	
Bunched Greens	32		90-95	Yes	Yes	No	Yes		Beets, Parsley, Radish, Spinach, Turnip
Cabbage, Chinese	32		95-100	No	No	No	Yes	2-3 months	
Cabbage, early	32	30.4	98-100	Yes	Yes	No	Yes	3-6 weeks	
Cabbage, late	32	30.4	98-100					5-6 months	
Cantaloupe	36-38		90-95	No	No	Medium	Yes		
Carrots, bunched	32		95-100	Yes	Yes	No	Yes	2 weeks	Ethylene may cause a bitter flavor
Carrots, immature	32	29.5	98-100					4-6 weeks	
Carrots, mature	32	29.5	98-100					7-9 months	
Cauliflower	32	30.6	95-98					3-4 weeks	
Cauliflower	32-35		90-95	No	No	No	Yes		
Celery	32	31.1	98-100	Yes	Yes	No	Yes	2-3 months	
Celeriac	32	30.3	97-99					6-8 months	
Chard	32		95-100					10-14 days	
Cherries	32-35		90-95	No	No	Very Low	No		
Cherries, sour	32	29.0	90-95					3-7 days	
Cherries, sweet	30-31	28.8	90-95					2-3 weeks	
Chicory	32-35		90-95	Yes	Yes	No	No		
Chicory, witloof	32		95-100					2-4 weeks	
Chinese pea pods	32-35		90-95	No	No	No	No		
Coconuts	55-60		80-85	No	No	No	No		Extended storage 32-35 F
Collards	32	30.6	95-100					10-14 days	
Corn, sweet	32	30.9	95-98	Yes	Yes	No	No	5-8 days	
Cranberries	38-42		90-95	Yes	No	No	No		
Cucumbers	50-55	31.1	95	No	No	Very Low	Yes	10-14 days	
Currants	31-32	30.2	90-95					1-4 weeks	
Eggplant	46-54	30.6	90-95	No	No	No	Yes	1 week	
Elderberries	31-32		90-95					1-2 weeks	
Endive	32	31.9	95-100	Yes	Yes	No	No	2-3 weeks	
Escarole	32-35		90-95	Yes	Yes	No	No		
Escarole	32	31.9	95-100					2-3 weeks	
Figs	32-35		90-95	No	No	Low	No		
Garlic	32	30.5	65-70	No	No	No	No	6-7 months	May be stored at 55-70 F for shorter periods
Ginger Root	60-65		65-70	No	No	No	No		
Gooseberries	31-32	30.0	90-95					3-4 weeks	
Grapefruit	55-60		90-95	No	No	Very Low	No		

Item	Temp (°F)		Humidity (%)					Storage	Notes
Grapes	31-32	29.7	85	No	No	Very Low	Yes	2-8 weeks	
Green Beans	40-45		90-95	No	No	No	Yes		
Green Peas	32-35		90-95	No	No	No	Yes		
Greens, leafy	32		95-100					10-14 days	
Guavas	45-50		90-95	No	No	Medium	Yes		
Herbs	32-35		90-95	No	Yes	No	Yes		
Horseradish	30-32	28.7	98-100					10-12 months	
Jicama	55-65		65-70					1-2 months	
Kale	32	31.1	95-100					2-3 weeks	
Kiwi, ripe	32-35		90-95	No	No	High	Yes		
Kiwi, unripe	32-35		90-95	No	No	Low	Yes, Very		
Kohlrabi	32	30.2	98-100	Yes	Yes	No	No	2-3 months	
Leeks	32	30.7	95-100	Yes	Yes	No	Yes	2-3 months	
Lemons	52-55		90-95	No	No	Very Low	No		
Lettuce	32	31.7	98-100	No	Yes	No	Yes	2-3 weeks	
Limes	48-55		90-95	No	No	Very Low	No		
Lychees	40-45		90-95	No	No	Very Low	No		
Mangos	50-55		85-95	No	No	Medium	Yes		
Melons, Casaba/Persian	50-55		85-95	No	No	Very Low	Yes		Riper melons may be stored at 45-50 F
Melons, Crenshaw	50-55		85-95	No	No	Low	Yes		Riper melons may be stored at 45-50 F
Melons, honey dew	50-55		85-95	No	No	Medium	Yes		Riper melons may be stored at 45-50 F
Mushrooms	32	30.4	95	No	Yes	No	Yes	3-4 days	
Napa	32-35		90-95	No	No	No	Yes		
Nectarines	31-32	30.4	90-95	No	No	High	No	2-4 weeks	
Okra	45-50	28.7	90-95	No	No	Very Low	Yes	7-10 days	
Onions	32-35		65-75	No	No	No	No		May be stored at 55-70 F for shorter period
Oranges	40-45		90-95	No	No	Very Low	No		
Papayas	50-55		85-95	No	No	Medium	Yes		
Parsley	32	30.0	95-100					2-3 months	
Parsnips	32	30.4	98-100	Yes	Yes	No	Yes	4-6 months	
Peaches	31-32	30.3	90-95	No	No	High	Yes	2-4 weeks	
Pears	29-31	29.2	90-95	No	No	High	Yes	2-7 months	
Peas, green	32	30.9	95-98					1-2 weeks	
Peas, southern	40-41		95					6-8 days	
Peppers, hot chili	32-50		60-70	No	No	No	Yes	6 months	
Peppers, sweet	45-55	30.7	90-95	No	No	No	No	2-3 weeks	
Persimmons	32-35		90-95	No	No	No	Yes, Very		
Pineapples	50-55		85-95	No	No	Very Low	No		Odor may influence avocados
Plums	31-32	30.5	90-95	No	No	High	Yes	2-5 weeks	
Pomegranates	41-50		90-95	No	No	No	No		
Potatoes	45-50		90-95	No	No	No	Yes		
Precut Fruit	32-36		90-95	No	No	Low	No		
Precut Vegetables	32-36		90-95	No	No	No	Yes		
Prunes	31-32	30.5	90-95	No	No	High	Yes	2-5 weeks	
Pumpkins	50-55	30.5	65-70	No	No	No	Yes	2-3 months	
Quinces	31-32	28.4	90	No	No	High	Yes	2-3 months	
Radishes, spring	32	30.7	95-100	Yes	Yes	No	Yes	3-4 weeks	
Radishes, winter	32		95-100					2-4 months	
Raspberries	31-32	30.0	90-95	No	No	Very Low	No	2-3 days	
Rhubarb	32	30.3	95-100	No	Yes	No	No	2-4 weeks	
Rutabagas	32	30.0	98-100	Yes	Yes	No	Yes	4-6 months	
Salad Mixes	32-35		90-95	No	Yes	No	Yes		
Salsify	32	30.0	95-98					2-4 months	
Spinach	32	31.5	95-100					10-14 days	
Sprouts	32-35		90-95	No	No	No	Yes		
Squashes, summer	41-50	31.1	95	No	No	No	Yes	1-2 weeks	
Squashes, winter	50	30.5	50-70	No	No	No	Yes	1-6 months	
Strawberries	32	30.6	90-95	No	No	Very Low	No	3-7 days	
Sweet Potatoes	55-60	29.7	85-90	No	No	No	Yes	4-7 months	
Tangerines	40-45		90-95	No	No	Very Low	No		
Tomatoes, mature green	55-70	31.0	90-95	No	No	Low	Yes	1-3 weeks	Ripening can be delayed by storing at 55-60 F
Tomatoes, ripe	55-70	31.1	90-95	No	No	Medium	No	4-7 days	
Turnip greens	32	31.7	95-100					10-14 days	
Turnips	32	30.1	95	Yes	Yes	No	Yes	4-5 months	
Watercress	32	31.4	95-100					2-3 weeks	
Watermelon	55-70		85-95	No	No	No	Yes, Very		Keep away from ethylene-producing fruits

INDEX

Information and Projects for the Self-Sufficient Homeowner

WHOLE HOME NEWS

From the Experts at Cool Springs Press and Voyageur Press

For even more information on improving your own home or homestead,
visit **www.wholehomenews.com** today! From raising vegetables to raising
roofs, it's the one-stop spot for sharing questions and getting answers
about the challenges of self-sufficient living.

- -

Brought to you by two publishing imprints of the Quayside Publishing Group, Voyageur
Press and Cool Springs Press, *Whole Home News* is a blog for people interested in the
same things we are: self-sufficiency, gardening, home improvement, country living,
and sustainability. Our mission is to provide you with information on the latest techniques and
trends from industry experts and everyday enthusiasts.

In addition to regular posts written by our volunteer in-house advisory committee, you'll also
meet others from the larger enthusiast community dedicated to "doing it for ourselves." Some
of these contributors include published authors of bestselling books, magazine and newspaper
journalists, freelance writers, media personalities, and industry experts. And you'll also find
features from ordinary folks who are equally passionate about these topics.

Join us at **www.wholehomenews.com** to keep the conversation going. You can also shoot us
an email at wholehomenews@quaysidepub.com. We look forward to seeing you online, and
thanks for reading!